AIChEMI

MODULAR INSTRUCTION

AF784839

Series A: PROCESS CONTROL

Volume 4: Advanced Control and Modeling Techniques

T. F. Edgar, Series Editor

AMERICAN INSTITUTE OF CHEMICAL ENGINEERS

© Copyright 1983

American Institute of Chemical Engineers
345 East 47th Street, New York, N.Y. 10017

Library of Congress Cataloging in Publication Data

(Revised for vol. 3)
Main entry under title:

American Institute of Chemical Engineers.
AIChEMI modular instruction: series A, process control.

Includes bibliographical references.
Contents: v. 1. Analysis of dynamic systems. v. 2. Feedback controller synthesis. v. 3. Design of sampled data (computer) control systems. v. 4. Advanced control and modeling techniques.
1. Chemical process control. I. Title II. Title: Process Control. III. Series: American Institute of Chemical Engineers. AIChEMI modular instruction; ser. 1, v. 1.
TP155.75.A45 1980 660.2′81 80-20826
ISBN 0-8169-0170-8 (v. 1)
ISBN 0-8169-0176-7 (v. 2)
ISBN 0-8169-0208-9 (v. 3)
ISBN 0-8169-0235-6 (v. 4)
ISSN 0270-6229

The appearance of the code below indicates the copyright owner's consent that, for a stated fee per article, copies of articles in this serial may be made for personal or internal use, or for the personal or internal use of specific clients. This consent is given on the condition that the copier pay the per-copy fee for each article (appearing as part of the code) through the Copyright Clearance Center, Inc., 21 Congress Street, Salem, Mass. 01970, for copying beyond that permitted by Sections 107 or 108 of the U.S. Copyright Law. This consent does not extend to copying for general distribution, for advertising or promotional purposes, for inclusion in a publication, or for resale.

ISSN 0270-6229/83/$3.75

CONTENTS

Solutions to the Homework Problems are available as a separate reprint from the AIChE Educational Services Dept., 345 East 46th St., New York, NY 10017. The cost is $5.00.

INTRODUCTION

In 1975 a new venture in education by and for the chemical engineering community was initiated. Prepared by the CACHE Corporation (Computer Aids for Chemical Engineering Education) and under the sponsorship of the National Science Foundation (Grant HES 75-03911), a series of small self-study fundamental concept modules for various areas of chemical engineering were commissioned, Chemical Engineering Modular Instruction, CHEMI.

It has been found in recent studies that modular study is more effective than traditional instruction in both university and continuing education settings. This is due in large measure to the discrete focus of each module, which allows the student to tailor the speed and order of his or her study. In addition, since the modules have different authors, each writing in his or her area of special expertise, they can be produced more quickly, and students may be asured of timely information. Finally, these modules have been tested in the classroom prior to their publication.

The educational effect of modular study is to reduce, in general, the number of hours required to teach a given subject; it is expected that the decreased time and expense involved in engineering education, when aided by modular instruction, will attract a larger number of students to engineering, including those who have not traditionally chosen engineering. For the practicing engineer, the modules are intended to enhance or broaden the skills he or she has already acquired, and to make available new fields of expertise.

The modules were designed with a variety of applications in mind. They may be pursued in a number of contexts: as outside study, special projects, entire university courses (credit or non-credit), review courses, or correspondence courses; and they may be studied in a variety of modes: as supplements to course work, as independent study, in continuing education programs, and in the traditional student/teacher mode.

A module was defined as a self-contained set of learning materials tat covers one or more topics. It should be sufficiently detailed that an outside evaluation could identify its educational objectives and determine a student's achievement of these objectives. A module should have the educational equivalent of a one to three hour lecture.

The CHEMI Project Staff included:

E. J. Henley, University of Houston, Director
W. Heenan, Texas A & I University, Assistant Director
Steering Committee:
L. B. Evans, Massachusetts Institute of Technology
G. J. Powers, Carnegie-Mellon University
E. J. Henley, University of Houston
D. M. Himmelblau, University of Texas at Austin
D. A. Mellichamp, University of California at Santa Barbara
R. E. C. Weaver, Tulane University
Editors:
Process Control: T. F. Edgar, University of Texas at Austin
Stagewise and Mass Transfer Operations: E. J. Henley, University of Houston, J. M. Calo, Brown University
Transport: R. J. Gordon, University of Flordia
Thermodynamics: B. M. Goodwin, Northeastern University
Kinetics: B. L. Crynes, Oklahoma State University
H. S. Fogler, University of Michigan
Material and Energy Balances: D. M. Himmelblau, University of Texas at Austin
Curriculum Analysis: E. J. Henley, University of Houston

The second phase of the project, designed to fill in gaps as well as develop new modules, is under the direction of D. M. Himmelblau, University of Texas at Austin.

Steering Committee:

B. Carnahan, University of Michigan
D. E. Griffith, Oklahoma State University
L. Harrisberger, University of Alabama
D. M. Himmelblau, University of Texas at Austin
V. Slamecka, Georgia Institute of Technology
R. Tinker, Technology Education Research Center

Editors (indicates a new task force head)*:

Process Control: T. F. Edgar, University of Texas at Austin
Stagewise and Mass Transfer Operations: J. M. Calo, Brown University
E. J. Henley, University of Houston
Transport: R. J. Gordon, University of Florida
Thermodynamics: G. A. Mansoori*, University of Ilinois at Chicago Circle
Kinetics: B. L. Crynes, Oklahoma State University
H. S. Fogler, University of Michigan
Material and Energy Balances: E. H. Snider*, University of Tulsa
Design of Equipment: J. C. Beckman, Arizona State University

Volume 1 of each series will appear in 1980; Volume 2 in 1981; and so forth. A tentative outline of all volumes to be produced in this series follows:

SERIES A: PROCESS CONTROL

Volume 1. Analysis of Dynamic Systems

A1.1	The Laplace Transformation	G. A. Atwood
A1.2	Linearization	H. E. Nuttall
A1.3	Time Domain Solutions of Linear, Time-Invariant Ordinary Differential Equations	G. S. Mueller
A1.4	First-Order Systems	V. E. Sater
A1.5	Second-Order Systems	V. E. Sater
A1.6	Higher-Order Systems	V. E. Sater
A1.7	Distributed Parameter Systems	P. R. Krishnaswamy
A1.8	Black-Box Process Modeling Using Process Reaction Curves	W. C. Clements, Jr.

Volume 2. Feedback Controller Synthesis

A2.1	Block Diagrams	D. W. Hubbard
A2.2	Open-Loop Frequency Response	W. C. Clements, Jr.
A2.3	Closed-Loop Frequency Response	L. M. Joseph
A2.4	Process Reaction Curves Controller Tuning Methods	G. A. Coulman
A2.5	Optimization of Closed-Loop Responses	G. Stephanopolous
A2.6	Nyquist Criterion	W. C. Clements, Jr.
A2.7	Routh Test for Stability	C. A. Smith
A2.8	Other Synthesis Methods	J. O. Hougen and T. F. Edgar

Volume 3. Design of Sampled Data (Computer) Control Systems

A3.1	Digital Computer Control of Chemical Processes	R. A. Mollenkamp
A3.2	Process Control Computer Equipment	R. A. Mollenkamp
A3.3	Z-Transforms	A. B. Corripio
A3.4	Z-Transform Diagram Analysis	A. B. Corripio
A3.5	Stability of Discrete Systems	A. B. Corripio
A3.6	Conversion of Continuous Time Models to Discrete Time Models	T. F. Edgar
A3.7	Sampling and Filtering for Discrete Time Systems	T. F. Edgar
A3.8	Digital Control Techniques	A. B. Corripio

Volume 4. Advanced Control and Modeling Techniques

A4.1	Pulse Testing of Dynamic Processes	G. H. Quentin
A4.2	Supervisory Control	R. Lavie
A4.3	Feedforward Control	T. F. Edgar and D. E. Seborg
A4.4	Cascade and Ratio Control	T. J. Ward
A4.5	Adaptive Control	T. F. Edgar and D. E. Seborg
A4.6	Deadtime Compensation	R. Lavie
A4.7	Interaction Matrix Analysis	T. J. Ward
A4.8	Selection of Control Measurements	M. E. Findley

Volume 5. Instrumentation and Control Applications

A5.1	Measurement Elements and Dynamics	D. E. Clough
A5.2	Final Control Elements and Dynamics	D. E. Clough
A5.3	Distillation Column Control	R. C. Waggoner
A5.4	Dynamics of Distillation	R. C. Waggoner
A5.5	Control of Chemical Reactors	W. L. Luyben
A5.6	An Overview of Application Control Design	E. H. Bristol

Publication and dissemination of these modules is under the direction of Harold I. Abramson, Staff Director, Educational Activities, AIChE. Technical Editor is Lori S. Roth. Chemical engineers in industry or academia who are interested in submitting modules for publication should direct them to H. I. Abramson, Staff Director, Educational Activities, American Institute of Chemical Engineers, 345 East 47th Street, New York, N.Y. 10017.

Pulse Testing of Dynamic Processes

G. H. Quentin

OBJECTIVES

At the completion of this module, given a physical system or process whose dynamic behavior is unknown, the student should be able to:

1. Conduct an experimental investigation of system dynamics using pulse testing methods.
2. Generate Fourier transforms and frequency response data from the experimental results using the computer program, PULSE.
3. Develop a linearized system model in the form of an *s*–domain transfer function from the frequency response data calculated from the experimental pulse testing results.

PREREQUISITE MATHEMATICAL SKILLS

The student should be able to solve:

1. Linear differential equations.
2. Complex algebra.
3. Laplace and Fourier transforms, and Fourier series.
4. System transfer functions.

PREREQUISITE ENGINEERING AND SCIENCE SKILLS

The student should be able to:

1. Model physical systems using mass, momentum, and energy balances.
2. Conduct a frequency response analysis using Bode diagrams.
3. Describe mathematically the linearized response of simple first and second-order lumped parameter systems to various forcing functions including step, ramp, and sinusoidal inputs.

To devise an effective process control scheme, an accurate model is needed for the dynamic behavior of the open–loop system. A simple model might be constructed based upon an a'priori theoretical analysis of the system; however, there are often serious discrepancies between models predicted from theory and the dynamic behavior of the actual system. To assume the role of an engineer responsible for process control, one must become familiar with experimental procedures for dynamic testing, which may be used either to validate analytical models or to formulate experimental models directly.

Pulse testing procedures are considered to be the most powerful approach to experimental determination of the dynamic behavior of a given physical system or process. This unit encourages the development of the ability to apply such pulse testing procedures. For a given physical system, one should be able: a) to conduct an experimental measurement of the dynamic response of the system following perturbation by a single input pulse, b) to translate the experimental results into a computer program (See Appendix) to generate Fourier transforms and frequency response data for the system, and c) to develop a system transfer function by analysis of the frequency response results.

PLANNING AN EXPERIMENTAL PROGRAM OF DYNAMIC TESTING

Dynamic testing requires that the system under study be perturbed slightly by a time–varying input function, such as a step change, a ramp function, or a sinusoidal wave, and subsequent measurement of the time–dependent (or transient) response of the system. From prior study, one should be familiar with the transient response of simple systems (i.e. lumped parameter, linear) of first and second order to these various forcing functions. What then are the advantages offered by pulse testing?

Frequency response tests have demonstrated several advantages over simple time–domain tests, such as those involving step change or ramp response. They are particularly useful in locating critical or resonant frequencies in a multiple–order system, that is, points of amplification or damping (i.e. dynamic gain) may be identified by frequency and related to specific system components. For example, introduction of the Electra turbo–prop aircraft several years ago was marred by a series of mysterious crashes, which frequency tests later showed to be the result of an unusual resonance involving wingtip vibration frequencies.

Frequency response testing with sinusoidal forcing however, is not only difficult from the standpoint of experimental accuracy, but is laborious and time-consuming. Moreover, when the system under study is an on–line manufacturing process, there is considerable hazard in the form of interrupted production or a deviation in product quality. How can the advantages of frequency response testing be achieved without these drawbacks?

Pulse testing can produce virtually the same results as a complete program of sinusoidal testing, and requires only a single brief experimental measurement of transient response to a pulse input with arbitrary form. Do not confuse this with an "impulse" or Dirac delta function, whose magnitude is infinite and duration is infinitesimal, but which cannot be readily achieved in practice. More realistically, pulse testing requires that process variables undergo only the slightest momentary departure from steady state, which in many instances may not be discernible on process instruments. Consequently, with the use of appropriate test instruments to observe the input and response, pulse tests can produce the desired experimental accuracy without needless interruption of quality or production.

It is important to achieve experimental accuracy in determining dynamic behavior from perturbations of small magnitude and brief duration. Test equipment must be selected to meet the requirements of precision and speed of response. Therefore, all sensors, signal conditioners, and recording equipment must have high gain characteristics for signal amplification and dynamic response characteristics significantly faster than the system under study. As a rule of thumb, the critical frequency of test equipment should be at least an order–of–magnitude higher than that of the fastest system component. (Caution must be exercised in the use of commercial process instruments, which may not meet these requirements.) Signal noise is another important consideration which can be a serious impediment, therefore proper signal shielding techniques must be observed when setting up test equipment to accomplish the highest signal-to-noise ratio. In a plant operating environment, special precautions must be taken because electrical interference can be severe.

CONDUCTING THE PULSE TEST

How can a whole series of sinusoidal inputs be replaced by a single pulse input and yet achieve similar results? The answer is that any transient wave shape or pulse form may be constructed from a summation of sine waves of varying amplitudes and frequencies.

Recall the Fourier series representation of a *periodic* pulse with arbitrary shape:

$$f(t) = A_o + \sum_{n=1}^{\infty} (A_n \cos n\omega_o t + B_n \sin n\omega_o t) \tag{1}$$

where, $\omega_o = 2\pi/P$, and P = period of the pulse. By simplifying and taking the limit as period P becomes infinite, one can derive the Fourier integral for a *non-periodic pulse*:

$$f(t) = \frac{1}{2\pi}\int_{-\infty}^{+\infty} F(j\omega)e^{j\omega t}d\omega \tag{2}$$

which also may be written as the Fourier transform, $F(j\omega)$:

$$F(j\omega) = \int_{0}^{+\infty} f(t)e^{-j\omega t}dt \tag{3}$$

Since the response of a linear system to a pulse input will also be a pulse, Equation 3 may be used to represent both the input and output response from a dynamic test as a Fourier series in the time domain. Consequently, pulse testing may be viewed as equivalent to measuring the dynamic response of a system to a series of sinusoidal inputs over a range of frequencies. What then is the implication of this view of pulse construction upon the results of a given pulse test?

Clearly, the range of useful frequency response results from a pulse test hinges primarily upon the shape and duration of the pulse. The input pulse must excite the system over a wide range of frequencies with amplitudes at high frequencies being distinguishable from noise. The ability of a given pulse to excite a system depends upon the number and amplitudes of the sinusoidal terms necessary to describe the pulse as a Fourier series.

The ability of a pulse to excite a system may be conveniently portrayed in terms of the "normalized frequency content," $N(\omega)$, which may be calculated by dividing the numerical value of the Fourier transform at a given frequency by the value of the transform at zero frequency:

$$N(\omega) = \frac{F(j\omega)}{F(j\omega)_{\omega=0}} = \frac{\int_0^P f(t)e^{-j\omega t}dt}{\int_0^P f(t)dt} = \frac{\text{Fourier Transform of Pulse Function, } f(t)}{\text{Area under Pulse Curve}} \tag{4}$$

Figure 1 is a comparison of the frequency content of several pulse forms, as a function of frequency, ω (rad/s), and pulse width, P(s).

Studies of pulse shape have indicated that certain regular shapes, e.g. weighted displaced cosine, will produce maximum system excitation; however, in practice these pulse shapes are not easily generated. By experience, it has proven sufficient to use pulse functions of an arbitrary form. By measuring the input variable, pulse width and height may be adjusted experimentally to give the most effective results. What are the practical constraints upon pulse width and height?

Figure 1 shows that shortening pulse width increases frequency content. However, if pulse width is too narrow for a given pulse height, or if the system is a sluggish one, the system excitation may not be distinguishable from noise. Increased pulse height will increase amplitudes at higher frequencies; there is, however, a constraint.

This unit attempts to construct a linear system model, as implied by the use of transfer functions. Real processes are generally non-linear, but linearity may be assumed in a small range about steady state. Therefore, one must not use pulse heights large enough to drive the system out of this linear range. As a check on linearity, it may be necessary to try pulses of similar shape and width but of varying height, including both positive and negative, to compare numerical values of the Fourier transform for the pulse function, which will be identical within the range of linearity.

A balance must be found between a sufficiently narrow or quick input pulse to provide reasonable system excitation and a small enough pulse height to maintain linearity and still provide distinguishable output amplitudes. A rule-of-thumb on pulse width is to keep it less than half the value of the smallest system time constant.

One realizes that the range and reliability of a system response depends largely upon the shape and size of the input pulse. However, there are several factors involving measurement of the output pulse which may also have some bearing upon experimental results.

One concern is truncation or shortening of the output pulse response. It has been shown that for lightly–damped systems, truncation of up to the last half of the output response will not seriously affect results. Since these particular system responses may take considerable time returning to steady-state, there is often an inclination to shorten them. However, in the case of heavily–damped systems, like those found

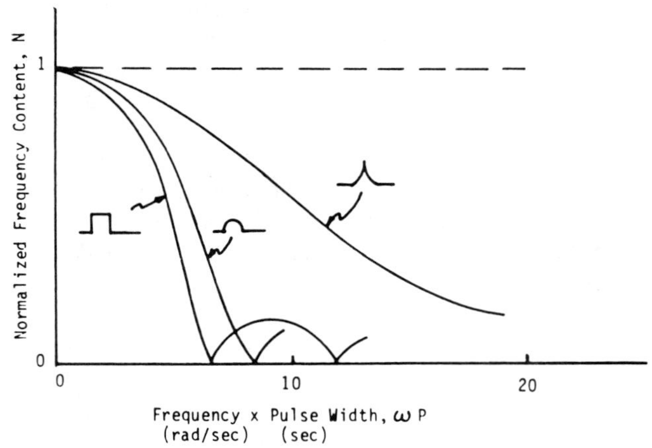

Figure 1.

so often in the process industries, premature truncation can distort the results by making a first–order frequency response appear like that of a second–order system. Since other errors may also occur, the general rule follows that the output pulse should be allowed to return to steady–state without truncation, unless it is altogether unavoidable.

Circumstances may arise where the output, and possibly the input pulse, do not "close" to their original steady–state value. This may indicate the presence of other factors causing a shift of the steady-state process operating level, especially when the response occurs over a length period of time. There is at least one example of pulse testing on a large-scale distillation column using an input pulse width of an hour or more, with a corresponding output response of a day or more in length. As a general rule, all pulses should be closed, however tests on such slowly-responding systems may take some time to complete making closure difficult.

In the event of non-closure of pulses, it is possible to use the resulting experimental data by making certain corrections to the mathematical algorithm used to compute the frequency response.

One should note, however, that failure to close the input pulse may greatly limit its frequency content. For example, an "open-ended" rectangular pulse approaches a step change which excites only low frequencies of a process. It has been shown that numerical differentiation of step response data can increase its frequency content, but there is also a loss of accuracy which makes it impractical. Although the Fourier transform normally does not apply to unbounded functions, e.g. a step-change or open-ended pulse, there are simple calculation methods to compensate for non-closure of pulses without numerical differentiation. The use of these techniques has been incorporated in a computer program to interpret pulse test data.

Noise and other random signal errors can also affect the reliability of the system response. While such distortion can impair the output pulse over the entire time span, it is most critical at the end where the return to steady state can be observed. The signal-to-noise factor is most important where discrete samples must be taken to represent a continuous pulse curve. Sampling intervals must be short enough to describe the signal accurately, and the tendency to "smooth" the data from discrete sampling should be avoided.

COMPUTATION OF FOURIER TRANSFORMS FROM EXPERIMENTAL RESULTS

Frequency response data for a given system may be computed from the ratio of the Fourier transform of the output pulse to that of the input pulse:

$$G(j\omega) = \frac{Y(j\omega)}{X(j\omega)} = \frac{\int_0^P y(t)e^{-j\omega t}dt}{\int_0^P x(t)e^{-j\omega t}dt} \tag{5}$$

where $x(t)$ = system input pulse function, $y(t)$ = system output pulse function, and the integral of each pulse function is zero for $P<t<\infty$.

Note that when $s=j\omega$, the Fourier transform becomes identical with the Laplace transform:

$$F(s)=\int_{o}^{\infty} f(t)e^{-st}dt \tag{6}$$

Then the Fourier transform ratio, $G(j\omega)$, becomes the system transfer function in the s–domain:

$$G(s)=\frac{\int_{o}^{P} y(t)e^{-st}dt}{\int_{o}^{P} x(t)e^{-st}dt} \tag{7}$$

The evaluation of the Fourier transform is further simplified by the use of the Euler identity:

$$e^{-j\omega t}=\cos \omega t-j \sin \omega t \tag{8}$$

from which:

$$F(j\omega)=\int f(t) \cos \omega t dt-j \int f(t) \sin \omega t dt \tag{9}$$

From this discussion, one should note that the Fourier transform may be expressed as a complex vector in the phase plane, as shown in Figures 2a and 2b. Note that the Fourier transform vector may take either form:

Rectangular Coordinates:

$$F(j\omega)=u+jv \tag{10}$$

or Polar Coordinates:

$$F(j\omega)=Re^{-j\phi} \tag{11}$$

where:

$$R=(u^2+v^2)^{1/2}$$
$$=|F(j\omega)| \text{ and, } \phi=\tan^{-1}(v/u)$$

The necessary frequency response data can then be calculated by taking the ratio of the magnitudes of the input and output transforms, and the difference of their phase angles, to generate the system transfer function in complex vector form:

Amplitude Ratio:

$$AR=|G(j\omega)|=\frac{|Y(j\omega)|}{|X(j\omega)|}=\frac{R_{out}}{R_{in}} \tag{12}$$

Phase Difference:

$$\phi_{\text{difference}}=\phi_y-\phi_x \tag{13}$$

Amplitude ratio and phase difference are, of course, functions of frequency and the value of the Fourier transform integral must be computed for each value of frequency over a wide range. This data then forms the basis for a Bode frequency response diagram displaying the dynamic characteristics of the system under study.

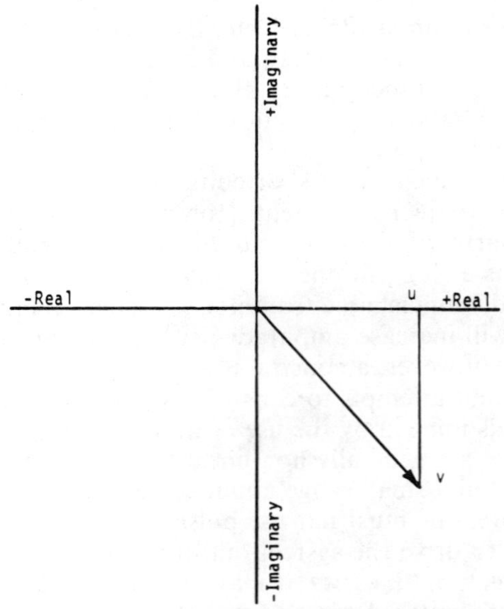

Figure 2a. Phase plane: rectangular coordinates.

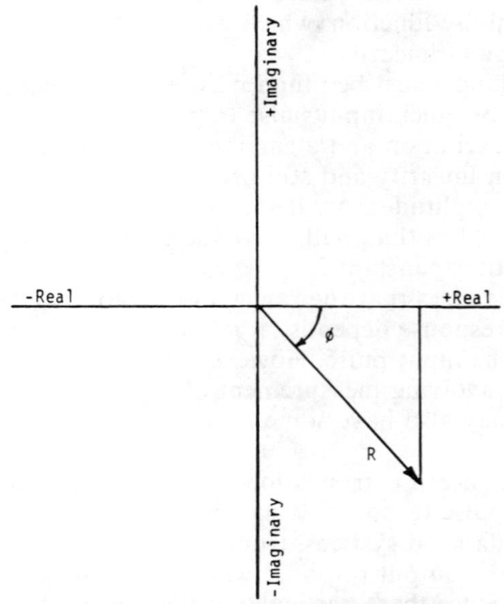

Figure 2b. Phase plane: polar coordinates.

The actual computer calculations require that the Fourier integral be approximated by a summation of a finite number of terms. Several forms of numerical approximation of the integral might be used, including rectangular, trapezoidal, or parabolic. The rectangular method is easier to use, but least accurate, while the parabolic is most accurate, but also more complex. The trapezoidal method, which is a compromise between rectangular and parabolic methods, would approximate the integral by a summation of terms representing the area of a trapezoid with base width, Δt, as follows:

$$F(j)=\int_{o}^{P} f(t)e^{-j\omega t}dt$$
$$=\left[\frac{\sin(\omega\Delta t/2)}{(\omega\Delta t/2)}\right]\sum_{k=1}^{k=n-1} f(k\Delta t)e^{-j\omega k\Delta t}\Delta t \tag{14}$$

The trapezoidal Fourier integral algorithm actually used to evaluate the transforms in the attached computer program (See Appendix) is much more complex than that shown here. It was originally developed by Hougen and Walsh; details of the algorithm are published elsewhere (*2*). The use of the computer program, PULSE, requires that each pulse curve function be entered as a series of data points with a specified time interval. The number of points used and the time interval chosen can affect the accuracy of the integral approximation, and therefore some criteria must be used as a basis for selection. It has been shown that the numerical approximation of the Fourier integral fails to converge at high frequencies, causing unreliable results for frequencies in that range. Lees and Dougherty (*3*) have shown that the limit of convergence of the various numerical approximations of the Fourier integral can be expressed as the following function of frequency, ω in radians/unit time, and sampling interval, Δt:

Convergence Limit:

$$\omega \Delta t \leqq \frac{\pi}{10} \qquad (15)$$

Earlier efforts had assumed this limit on reliable frequency range to be at least a decade higher. Consequently, a large number of data points and small time interval is necessary to ensure accurate data. This selection criteria should be used to determine the maximum time interval when digitizing a continuous data record or setting the sampling the sampling interval when a continuous signal is not available.

For additional convenience, the output from the computer program, PULSE, includes an indication of the frequency at which the normalized frequency content of the input pulse falls below 0.01. Various error criteria indicate that reliability is actually lost well before this, approximately at the point where the normalized frequency content falls below 0.3.

The treatment of open-ended pulses in the computer program, PULSE, is based upon the following function:

$$G(s) = \frac{e^{-sT_2} y(T_2) + s \int_o^{T_2} y(t) e^{-st} dt}{e^{-sT_1} x(T_1) + s \int_o^{T_1} x(t) e^{-st} dt}$$

where $s = j\omega$ (16)

and the points T_1 and T_2 are well beyond the point where the time histories of the pulses $x(t)$ and $y(t)$ have returned to steady state (Figures 3a and 3b). Therefore, it is sufficient to enter pulse time-histories up to the point of nonclosure, P, and to signify the pulse open-ended for application of the correction algorithm.

A pulse may very nearly approach closure, but retain a permanent steady-state deviation due to some external factor, such as test instrument zero drift. In this case, the computer program permits an alternate

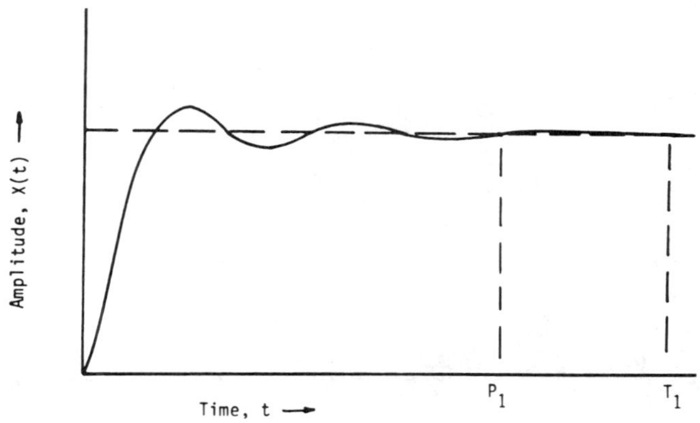

Figure 3a.

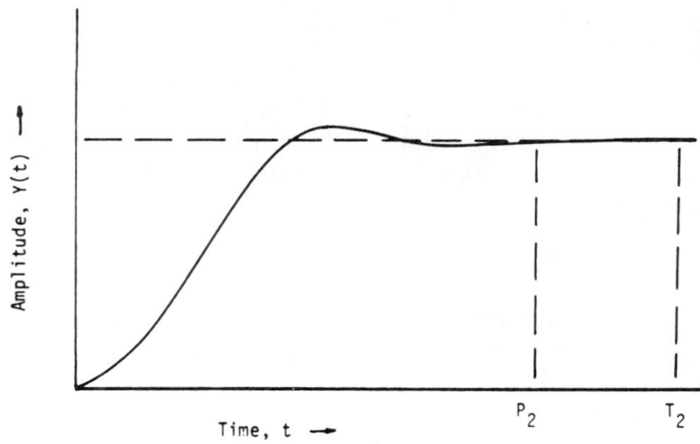

Figure 3b.

closure technique in which a linear base line is drawn from the origin through the point of nonclosure (Figure 4).

DEVELOPING THE SYSTEM TRANSFER FUNCTION MODEL FROM FREQUENCY RESPONSE DATA

The frequency response data generated by the PULSE program may be conveniently plotted on a Bode diagram, with the Log Amplitude Ratio and Phase Difference versus Log Frequency. The Bode diagram may be used to deduce the system model, expressed as an s-domain transfer function.

If a linear system model is available beforehand, the experimental frequency response results may be used to validate the predicted model. In this event, it is a simple matter to plot both the theoretically predicted frequency response and the experimental frequency response data on the same Bode diagram. Techniques for use of the Bode diagram have been introduced in an earlier module in this series.

The more difficult task is to construct a transfer function model directly from the Bode diagram representation of the experimental results without an analytical model for comparison. The general method to accomplish this requires the fitting of standard

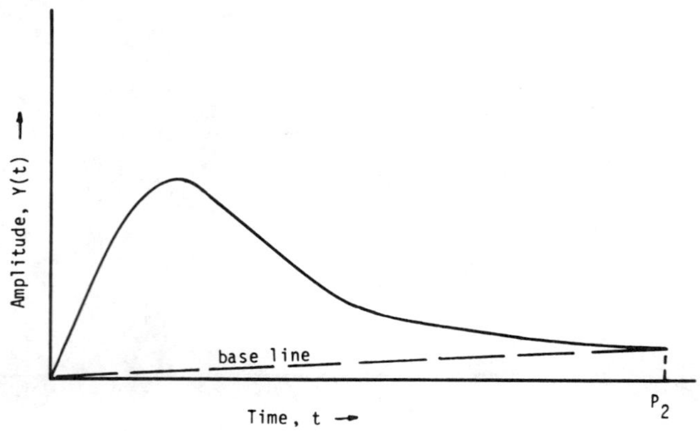

Figure 4.

function profiles to the experimental results on the Bode diagram, and constructing an overall model as a composite of the individual *s*-domain functions.

From prior study, one should already be familiar with the necessary *s*-domain functions and readily identify their profiles on a Bode diagram; they include:

a) pure integration (lag function)
b) pure differentiation (lead function)
c) first-order system, lead or lag
d) second-order system, under-, over-, or critically-damped
e) pure delay time

Among the dynamic parameters that will require evaluation on the Bode diagram in order to derive the system model are the following:

a) steady-state gain, or zero-frequency amplitude ration
b) corner frequency of a first-order system
c) damping ratio of a second-order system
d) resonant frequency of either second-order system or a distributed parameter system

It may be helpful to make templates of amplitude ratio (AR) and phase difference (PD) profiles for a first-order system and second-order systems with various damping ratios (i.e. critical and under-damped), using a standard size semi-log paper and a convenient material for the templates. A detailed procedure for fitting such profiles to Bode diagrams is given elsewhere (*2*), together with a worked example of an over-damped, second-order system.

Basically, there are a few simple guidelines to aid this exercise, the construction of an unknown transfer function. They include:

(1) Fit the simplest function possible. A good fit must match both the AR and PD plots.
(2) If a given profile fits the AR plot well, but not the PD plot, try adding dead time. Recall that PD decreases rapidly with frequency.
(3) Second-order, under-damped systems may be easily identified by their resonant peak, when the damping ratio is low. The height of the peak will help identify the damping ratio, but recall that the PD characteristics also vary with this ratio.
(4) A series of resonant peaks with valleys in between them indicates harmonic frequencies; these are found in distributed systems such as transmission lines or heat exchangers.
(5) If the AR plot is not fit well by a first, second, or *n*th–order profile, but the slope appears to be of intermediate value, try adding combinations of first and second-order systems (or others) with different corner frequencies until AR and PD both give reasonable fit. For example, adding two first-order profiles with different corner frequencies will provide a "transition" slope (i.e. less than -2 and greater than -1) in the region between the corner frequencies. The rate of transition from a slope of -1 to a slope of -2 is, of course, a function of the amount of separation between corner frequencies.

In summarizing this procedure, one may be in the position of essentially trying to fit the best "second-order plus dead-time" system that one can. Do not be dismayed if the system under study does not turn out to be complex, because many systems in the process industries contain large capacitances and resistances which cause a damping effect, negating the effects of high frequency variations. This is noted on the Bode diagram where high frequency oscillations in a sluggish system will appear far to the right on the diagram where attenuation is great and their effect is nil. If a good fit is achieved with a linearized system model from a pulse test which excited the system over a wide range of frequencies, remember that the model is valid at that particular process operating point. Consider conducting additional tests over a range of operating conditions, and possibly with other input variables which might have an effect upon the system.

LITERATURE CITED

1. Coughanowr, D. R. and L. B. Koppel, "Process Systems Analysis and Control," Chapter 24, McGraw-Hill, New York (1965).
2. Hougen, J. O. "Measurements and Control Applications for Practicing Engineers," Chapters 7 and 8, Appendix 3, Barnes & Noble, Boston (1972).
3. Lees, S. and R. C. Dougherty, "Numerical Fourier Transform Calculations for Pulse Testing Procedures," *Journal of Basic Engr., Transactions of ASME*, p. 445, June (1967).
4. Luyben, W. L. "Process Modeling, Simulation, and Control for Chemical Engineers," Chapter 9, McGraw-Hill, New York (1973).
5. Pollock, G. G. and A. I. Johnson, "Comparison of Dynamic Testing Methods and their Application to an Extraction Process," *Can. J. Chem. Engr.*, v. 47, p. 565, Dec. (1969).
6. Tyner, M. and F. P. May, "Process Engineering Control," Appendix F, Ronald Press, New York (1968).

STUDY PROBLEMS

1. Give several reasons why pulse testing is preferred for dynamic testing programs.

2. What is the significance of the "frequency content" of an input pulse?

3. Give the reasons for constraints on both the height and width of the input pulse.

4. Is it necessary to have a continuously recorded system output pulse response? If not, give any qualifications.

5. Can pulse testing be applied to non-linear systems? If yes, give any qualifications.

6. Suppose one anticipates a second-order system response from a pulse test, but the system is expected to be very slow in responding to a pulse input. If damping ratio is high, should one truncate the output pulse to reduce the amount of time required for the test?

7. What steps can be taken in conducting a pulse test if one encounters signal noise and/or random error and it affects the frequency response results?

8. Can one use experimental data if an output pulse fails to close? If so, explain how.

9. Would the presence of pure delay (dead-time) in a system cause erroneous pulse test results? How would it manifest itself on a Bode diagram? How could one compensate for it?

10. What could one deduce about a system model and its corresponding test results if the output pulse and input pulse forms turned out to be nearly identical? What is indicated if the output pulse had the same form, but appeared some time after the input pulse?

HOMEWORK PROBLEMS

1. Use the computer program, PULSE (See Appendix), to generate frequency response data for the experimental results of a pulse test on an open-loop process (Table 1). Plot the results on a Bode diagram.

2. From the computer output in Homework Problem 1, use the data on the input pulse transform vector magnitude to plot the normalized frequency content of the input pulse versus frequency. (Note: input transform vector magnitude may be divided by its zero frequency magnitude to yield normalized frequency content.)

3. Deduce the system transfer function from the Bode diagram results plotted in Homework Problem 1.

4. A control system has been added to the open-loop process in Problem 1. A pulse test has been made, using a set point change as the input, with a rectangular pulse input 500.0 units high and 1 second in duration. The system was tested while under closed-loop proportional control. The output pulse response of the closed–loop system is given in Table 2.
Use the computer program, PULSE, to generate frequency response data, and plot these results on a Bode diagram. Does the plot produce the anticipated results when compared with that in Problem 1?

5. The proportional controller in Homework Problem 4 is replaced with a 2-mode proportional-integral controller and the pulse test repeated with the closed–loop system and a set point change. See Table 3. The rectangular pulse input has a magnitude of 500.0 units, but the time interval is now 1.4 seconds. Use the PULSE program to generate frequency response results and plot a Bode diagram. Are the results as anticipated?

6. If you have access to an analog or hybrid computer, program a second-order under-damped system with critical frequency of approximately 1 cycle/s and a damping ratio of 0.4. Using the analog computer simulation, conduct an experimental pulse test by means of the manual switch input, plotting both input and output pulses. Then use the computer program, PULSE, to generate experimental frequency response data. Compare your experimental results with the known form of the second-order response profile on a Bode diagram.

Table 1.

Time (s)		Input Pulse, $x(t)$	Time (s)	Output Pulse, $y(t)$
0.0	(Rectangular	500.0	2.4	150.0
1.0	Pulse Form)	500.0	2.8	120.0
			3.2	100.0
Time (s)		Output Pulse, $y(t)$	3.6	75.0
0.0		0.0	4.0	60.0
0.4		30.0	4.4	45.0
0.8		90.0	4.8	35.0
1.2		160.0	5.2	25.0
1.6		180.0	5.6	20.0
2.0		170.0	6.0	15.0
			10.0	0.0

Table 2.

Time (s)	Output Pulse, $y(t)$	Time (s)	Output Pulse, $y(t)$
0.0	0.0	2.0	−20.0
0.4	40.0	2.4	−20.0
0.8	65.0	2.8	5.0
1.2	60.0	3.2	5.0
1.6	10.0	4.8	0.0

Table 3.

Time (s)	Output Pulse, $y(t)$	Time (s)	Output Pulse, $y(t)$
0.0	0.0	3.2	+10.0
0.4	30.0	3.6	10.0
0.8	65.0	4.0	0.0
1.2	40.0	4.4	−10.0
1.6	−20.0	4.8	−10.0
2.0	−60.0	5.2	−5.0
2.4	−50.0	8.0	0.0
2.8	−10.0		

APPENDIX—COMPUTER PROGRAM

Program Name: PULSE
Class: Normal FORTRAN
Type: Self-Standing

Title: Pulse Test Frequency Response Analysis

Abstract:

The Program produces a frequency response analysis of the dynamic behavior of a given physical system. Program input consists of results from time-domain experiments using pulse testing procedures. Input is in the form of magnitude versus time data for each system input pulse and each output response. The Fourier transform is computed for each system input and output by a subroutine using a trigonometric Fourier integral algorithm. These transforms are calculated over a range of frequencies as given in the program or optionally specified by the user. The frequency response for each specified system input-output pair is then computed from the ratio of Fourier transforms for that pair. The Fourier transform of each system input and output is printed in the form of phase-plane vectors in both rectangular and polar notation, at each selected frequency. Frequency response results are printed in the form of amplitude ratio and phase difference for each specified system input-output pair, at each selected frequency.

Related Programs: Bode Diagram Plotting

If on-line digital computer plotting capability is available, the frequency response results may be displayed in the form of a Bode diagram showing amplitude ratio and phase difference plotted against frequency on a logarithmic scale. An appropriate plotting subroutine may be added.

Program Development:

Computer Used: IBM 360/75
Minimum Core: 200 k bytes
Operating Environment: Batch
Language: FORTRAN IV
Typical Run Time: 30 seconds
Number of Source Cards: 520

USER INSTRUCTIONS FOR PROGRAM PULSE

Program input data is broken down into a series of categories called "lists," which make up the complete data deck for each run. These are shown in the table below:

Table 4.

List Number	List Function
1	Initialization and alphanumeric title to identify output
2	Alphanumeric title to identify output, used with rerun option
3	Program output alternatives, including specification of transform pairs for transfer function
4	Optional frequency range selection
5	Optional frequency points selection
6	Pulse data input, and alphanumeric pulse identification (up to 9 separate curves may be entered with separate list for each)
90	Data changes for optimal rerun of same case
99	End of data for each case, signals program run

The following is a series of special instructions for use of the various data lists given above:

(1) The first card for each list will have the list number in 2I1 format in the first two columns on the card.

(2) For lists with numerical data entries to follow (i.e. list nos. 3, 4, 5, 6, and 90), the numerical data card input format field shall be specified on the first card for that list, beginning in column 7. It shall apply to numerical data in that list only. Title formats need not be specified.

(3) A separate list 6 entry shall be included for each separate pulse curve.

(4) Lists 1, 2, and each list 6 entry shall *always* have a second card with alphanumeric title information for printer output identification. A 60 character field, beginning in column 1, is used for lists 1 and 2; with a 36 character field, beginning in column 1, used for each list 6 entry. Numerical cards begin with card 3, i.e. after the title card.

(5) Lists 3, 4, 5, and 90 have no alphanumeric title entries, therefore numerical data entries begin with card 2 of each list.

(6) Note that a case may be rerun by entering only those items which need to be changed. Specific data items may be changed by use of list 90, identifying changes by their location in the "DB" array. This may be useful where a data list is extensive. In the case of short lists (e.g. list 3), it may prove as easy to merely re-enter the complete original list with changes.

(7) List 2 is used in reruns of a case to provide output identification. Remember not to use list 1 for this purpose, since it will initialize the program, clearing out old data.

(8) Experimental data from a series of pulse tests, involving up to 9 separate pulse curves, may all be entered at one time with the first case entered. This may be important where a single input pulse results in several different output responses simultaneously from different points in the test system. However, a separate case must be run for each transform pair specified for a system transfer function, since list 3 specifies only one pair at a time and must be changed for each case.

Data entries for each of the data input lists in the program are identified in Table 5, and each is related to a location in the "DB" array which is the main data storage bank in the program. Examples of card sequences for each list are also given. Figure 5 is the program, PULSE.

Table 5.

CARD SEQUENCE EXAMPLE FOR LISTS 1 AND 2 (ALPHANUMERIC TITLE ENTRIES)

column no. →	1 →	→ 60
card 1	1	
card 2	PULSE TEST PROGRAM MAIN TITLE	
card 3	2	
card 4	PULSE TEST CASE SUB-TITLE	

LIST 3 PROGRAM OUTPUT ALTERNATIVES

Data Item No. (location in DB array)	Sequence of Data Entries (all appear on card 2 of list 3)
1	Pulse number (k) to be used as system input in transfer function
2	Pulse number (k) to be used as system output in transfer function
3	Amplitude normalization code (0 = Normalized, 1 = Not Normalized)
4	Transform Print Code (0 = Print Both Transforms, 1 = Input Transform Only, 2 = Output Transform Only, 3 = No Transforms Printed)
5	(To be used for Bode Plot Code if a plotter routine is available)

CARD SEQUENCE EXAMPLE FOR LIST 3 (DATA ENTRIES)

column no. →	1	7	11	16	21	26	31
card 1	3	(6X,5G5.2)					
card 2			1	2	0.0	0.0	0.0

LIST 4 OPTIONAL FREQUENCY RANGE SELECTION

Data Item No. (location in DB array)	Sequence of Data Entries (all appear on card 2 of list 4)	Default Value
6	Frequency Units (0 = Radians/Time, 1 = Cycles/Time)	0
7	Number of Frequency Decades (maximum of 5)	5
8	Number of Frequencies per Decade (maximum of 10)	10
9	Initial Frequency (or lowest value)	0.001

CARD SEQUENCE EXAMPLE FOR LIST 4 (DATA ENTRIES)

column no. →	1	7	11	16	21	26
card 1	4	(6X,4G5.2)				
card 2			1	1	10	0.1

(continued)

LIST 5–OPTIONAL FREQUENCY POINT SELECTION

Data Item No. (Location in DB Array)	Sequence of Data Entries (beginning with card 2 of list 5)	Default Value
6	Frequency Units (0 = Radians/Time, 1 = Cycles/Time)	0
10	Number of Frequencies to be entered (max. of 50)	50
101 to 150	Actual Frequency Values Entered	0.001 to 1.0 (evenly-spaced on log plot)

CARD SEQUENCE EXAMPLE FOR LIST 5 (DATA ENTRIES)

column no.	1	5	7	10	20	30	40	50		
card 1	5	(2G5.2/10G5.2)								
card 2	1	10								
card 3	0.1	0.2	0.3	0.4	0.5	0.6	0.7	0.8	0.9	1.0

LIST 6–PULSE DATA INPUT

Data Item No. (Location in DB Array)	Sequence of Data Entries (beginning with card 3 of list 6)
10k + 1	Pulse Number, k where 1 ⩽ k ⩽ 9
10k + 2	Pulse Height prior to Excitation, at time, t = 10−
10k + 3	Pulse Height at Final Steady-State, at time, t = ∞
10k + 4	Pulse Height Units Conversion Factor (enter 1.0 for no conversion)
10k + 5	Pulse Height Data Output Code (0 = No Print-out of Pulse Height Data, 1 = Pulse Height Data Printed)
10k + 6	Pulse Closure Correction Code (0 = No Correction, 1 = Linear Base Line Correction, 2 = Open Pulse Correction Algorithm)
10k + 7	Number of Curve Sections, M, with different time intervals
((10k+200) + m)	Time Interval for Initial Section of Curve (m = 1)
((10k+300) + m)	Number of Last Point in Initial Section of Curve (m = 1)
(REPEAT THE ABOVE PAIR OF ENTRIES FOR EACH CURVE SECTION, m = 1 to M)	
((10k+200) + M)	Time Interval in Final Section of Curve (m = M)
((10k+300) + M)	Number of Last Point in Final Section of Curve (m = M)
10k + 8	Total Number of Points in Entire Curve, N, where 2 ⩽ N ⩽ 1000
{1000k + 1 to 1000k + N}	Pulse Height Data starting at time, t = 0+

CARD SEQUENCE EXAMPLE FOR LIST 6 (TITLE AND DATA ENTRIES)

column no.	1	7	10	20	30	40	50	60				
card 1	6	(2(10G5.2))										
card 2	INPUT PULSE NAME											
card 3	1	0.0	0.0	1.0	1.0	0	1	0.1	10	10		
card 4	1.0	2.0	4.0	8.0	10.0	9.0	7.0	5.0	3.0	1.0		
(REPEAT THE SEQUENCE FOR EACH OUTPUT PULSE CURVE)												
card 5	6	(12G5.2/10G5.2)										
card 6	OUTPUT PULSE NAME											
card 7	2	0.0	0.0	1.0	1.0	0.0	2	0.1	6	0.5	10	10
card 8	0.1	0.2	0.3	0.4	0.35	0.3	0.2	0.15	0.1	0.05		
(REPEAT THE SEQUENCE, AS NECESSARY, FOR EACH ADDITIONAL PULSE CURVE)												

LIST 90 DATA CHANGES FOR OPTIONAL RERUN OF SAME CASE

Position of Data Entry in this list only	Sequence of Data Entries (beginning with card 2 in list 90)
1	Number of Entries in DB Array to be changed, Nch
2	First Item to be changed, by location in DB Array
3	Numerical Value to be entered for first item changed
(REPEAT THE ABOVE PAIR OF ENTRIES FOR EACH ITEM TO BE CHANGED, n = 1 to Nch)	
2 + Nch	Last Item to be changed, by location in DB Array
3 + Nch	Numerical Value to be entered for last item changed

(continued)

CARD SEQUENCE EXAMPLE FOR LIST 90 & LIST 99 (DATA ENTRIES)

column no.	1		7						
card 1	90		(9G5.2)						
card 2		4	1001	0.0	1010	0.0	2001	0.0	2010	0.0
card 3	99								

(Explanation: The above cards would make 4 data item corrections from those entered in the list 6 example above. The first and last points in both the input and output pulse curves, each with 10 total points, would be adjusted to a value of 0.0. The points are located by their position in the DB array, i.e. 1001 & 1010 for the first and last points in the input curve and 2001 & 2010 for the first and last points in the output curve.)

```
C       PROGRAM NAME =  PULSE
C
C       THIS PROGRAM COMPUTES FOURIER TRANSFORMS FOR CLOSED OR OPEN PULSES
C       TRANSFER FUNCTIONS FOR PAIRS OF CURVES MAY ALSO BE CALCULATED.
C       DEFINITION SECTION
C       A      FOURIER TRANSFORM(FT) AMPLITUDE
C       AI     FT AMPLITUDE OF INPUT IMPULSE
C       AIN ALPHANUMERIC WORD
C       AR  AMPLITUDE RATIO (OUTPUT/INPUT)
C       APL  ALOG10(AR)
C       AZ   AREA UNDR PULSE
C       AZ AREA UNDER PULSE (FT AMPLITUDE AT ZERO FREQUENCY)
C       AZI   AREA UNDER INPUT PULSE
C       BLANK ALPAMERIC WORD
C       CF  CHART FACTOR
C       CFI  CHART FACTOR FOR INPUT PULSE
C       DB     INPUT DATA BASE ARRAY
C       F     FREQUENCY  ARRAY(CYCLES/TIME IF KFU=1, ELSE RAD/TIME)
C       FI INITIAL FREQUENCY
C       FLM    FREQUENCY LIMIT BASED ON STEP SIZE
C       FM FREQUENCY ARRAY FOR ONE DECADE
C       FU     ALPHAMERIC WORD FOR FREQUENCY UNITS
C       GAIN  STEADY STATE GAIN (OUTPUT PULSE AREA / INPUT PULSE AREA)
C       H TIME BETWEEN POINTS IN J-TH SECTION OF K-TH PULSE
C       HM  MINIMUM TIME INTERVAL
C       HMI    MINIMUM TIME INTERVAL FOR INPUT PULSE
C       IA     NUMBER OF THE FIRST POINT IN A PULSE SECTION
C       IB     NUMBER OF THE LAST POINT IN A PULSE SECTION
C       JP     NUMBER OF POINTS FOR PLOT
C       K     PULSE NUMBER
C       KBP     BODE PLOT CODE
C       KEE     PULSE CLOSURE CODE
C       KEEI PULSE CLOSURE CODE FOR INPUT PULSE
C       KFD     FREQUENCY DATA TYPE CODE
C       KFL  FREQUENCY LIMIT PRINT SWITCH
C       KFU     FREQUENCY UNITS CODE
C        KRN     RESPONSE NORMALIZATION CODE
C       KTP     TRANSFORM DATA PRINT CODE
C       NB  NUMBER OF LAST POINT IN J-TH SECTION OF A PULSE
C       NBK NUMBER OF LAST POINT IN J-TH SECTION OF K-TH PULSE
C       NCI    PULSE NUMBER OF INPUT CURVE
C       NCO    PULSE NUMBER OF OUTPUT CURVE
C       ND    NUMBER OF DECADES
C       NF    NUMBER OF FREQUENCIES
C       NFD    NUMBER OF FREQUENCIES PER DECADE
C       NFP1    NF+1
C        NP    NUMBER OF DATA POINTS
C       NS     NUMBER OF PULSE CURVE SECTIONS
C       OUT    ALPHAMERIC WORD
C        P    FT PHASE ANGLE
C       PA     ALPHAMERIC ARRAY FOR PULSE ID
C       PD  PULSE DATA ARRAY
C       PH     PULSE HEIGHT DATA ARRAY
C       PI    FT PHASE ANGLE FOR INPUT IMPULSE
C       PL     PHASE LAG(OUTPUT-INPUT)
C       R   FT REAL PART
C       RI  FT REAL PART FOR INPUT PULSE
C       T   TIME
C       TT  TOTAL TIME
C       TTI  TOTAL TIME FOR INPUT IMPULSE
C       TITLE ALPHAMERIC ARRAY FOR PLOT(LIST 2)
C       X   LOG FREQUENCY ARRAY FOR PLOT
C       Y   PULSE HEIGHT ARRAY
C       Y1  LOG AMPLITUDE ARRAY FOR PLOT
C       Y2 PHASE LAG ARRAY FOR PLOT
C       YF  PULSE HEIGHT AT INFINITE TIME
C       YZ  PULSE HEIGHT AT TIME 0-
C        Z  FT IMAGINARY PART
C       ZI     FT IMAGINARY PART FOR INPUT IMPULSE
C
C       NONCE VARIABLES - A1,A2,AK,AN,ANFD,I1,I2,J1
C
C       DECLARATIVE SECTION
        DIMENSION FMT(15)
        DIMENSION   DB(10000),TTL(30),TITLE(12),A1(9)
        DIMENSION F(52),FM(10),H(10,9),PA(9,9),NBK(10,9),
       1PD(10,9),PH(1000,9),R(52),RI(52),X(51),
       2  Y(1000),Y1(51),Y2(51),Z(52),ZI(52),NB(10)
        EQUIVALENCE  (DB(11),PD),(DB(211),H),  (DB(311),NBK),
       1    (DB(1001),PH)
        DATA  BLANK    /4H    /
        DATA  AIN/4H  IN/,OUT/4H OUT/,CYC/4HCYC./, RAD/4HRAD./
        DATA    KPLOT /0/
C
C       FILE SECTION
  181 FORMAT(/16X,    16H    PULSE NO. = , I1, 2X, 9A4)
  182 FORMAT(/24X, A4, 16HPUT PULSE NO. = ,I1, 2X, 9A4)
  1830FORMAT(10X,37HORDINATES OF CURVE WERE MULTIPLIED BY F8.3,22H BEFOR
     1E TRANSFORMATION/10X,17HTOTAL TIME SPAN = F11.4,4X,62HAREA WITHIN
     4THIS TIME SPAN (AFTER SUBTRACTING INITIAL VALUE) = F12.4)
  1840FORMAT( 10X,70HLINEAR CORRECTION WAS MADE TO FORCE FINAL VALUE TO
     4EQUAL INITIAL VALUE)
  185 FORMAT( 10X,42HFINAL VALUE ASSUMED TO PERSIST TO INFINITY)
  186 FORMAT(/20X, 50HFREQUENCY     REAL PART      IMAG. PART      AMPLI
     1            , 19HTUDE     PHASE ANGLE/
     2          20X, A4,  5H/TIME, 47X, 15HDEGREES (0-360D)
  187 FORMAT( 18X, 4(1PE11.4,4X),0PF8.2)
  189 FORMAT(/10X, 41HFREQUENCY LIMIT BASED ON INTERVAL SIZE = ,1PE10.3)
  190 FORMAT( 10X, 20HSTEADY STATE GAIN = , F12.4)
  191 FORMAT(/ 2X, 40H   FREQUENCY RESPONSE (OUTPUT/INPUT)      ,
     1      40H FOURIER TRANSFORM OF THE  INPUT CURVE   ,
     2       38H FOURIEP TRANSFORM OF THE OUTPUT CURVE/
     3   2X, 40HFREQUENCY  AMPLITUDE   LOG      PH. DIF
     4      40HREAL PART IMAG. PART  AMPLITUDE PH.ANG   ,
     5      38HREAL PART IMAG. PART  AMPLITUDE PH.ANG/
     6    2X,A4, 36H/TIME     RATIO   AMP. RAT  DEGREES   ,
     7         33X, 4HDEG., 36X, 4HDEG.)
  192 FORMAT(/ 2X, 40H   FREQUENCY RESPONSE (OUTPUT/INPUT)      ,
     1      25H FOURIER TRANSFORM OF THE, A4, 11HPUT CURVE   /
     2   2X, 40HFREQUENCY  AMPLITUDE   LOG      PH. DIF   ,
     3         40HREAL PART IMAG. PART  AMPLITUDE PH.ANG   /
     4        2X, A4, 36H/TIME     RATIO   AMP. RAT  DEGREES   ,
     5          33X, 4HDEG.)
  193 FORMAT(/ 2X, 40H   FREQUENCY RESPONSE (OUTPUT/INPUT)      / ,
     1          2X, 40HFREQUENCY  AMPLITUDE   LOG      PH. DIF  /
     2             2X, A4, 36H/TIME     RATIO   AMP. RAT  DEGREES  )
  194 FORMAT(  1X, 50HINPUT AMPLITUDE IS LESS THAN ONE PER CENT OF PULSE
     1       , 50H AREA. RESULTS AT HIGHER FREQUENCIES ARE QUESTIONA
     2       ,  4HBLE.)
  195 FORMAT(2(1X,1PE10.3),0PF9.5,F9.2,2(3(1X,1PE10.3),0PF7.2))
  901 FORMAT( 20A4)
  902 FORMAT(    18A4)
  903 FORMAT(  47H1FREQUENCY RESPONSE ANALYSIS OF PULSE TEST DATA,28X,
     1      // 1X, 15A4 / 1X, 15A4)
  904 FORMAT(//30X, 18HINPUT DATA SUMMARY,
     1       // 1X, 44H ITEM  VALUE       ITEM  VALUF       ITEM  VALUE,
     2        4X, 28H ITEM  VALUE      ITEM  VALUE)
  905 FORMAT(    1X,I4,1PG12.4, I4,G12.4, I4,G12.4, I4,G12.4,I4,G12.4)
  906 FORMAT(/)
  907 FORMAT(//38H JOB TERMINATED - BAD DATA BEFORE CARD // 1X, 20A4)
  908 FORMAT(7X,33H SUCCESSFULLY READ LIST NUMBER = ,I2,7H LNB = ,I2/
     1  7X,7H FMT = ,15A4,/)
  909 FORMAT(I1,I1,4X,15A4)
  910 FORMAT(    9A4)
  911 FORMAT(7X,43H ERROR ENCOUNTERED IN READ OPERATION STMT 5/)
  912 FORMAT(7X,43H ERROR ENCOUNTERED READING DATA IN LIST NO.,I2/)
  913 FORMAT(6X,36H IMPROPER LIST 9 BRANCH......LNB = ,I1/)
  914 FORMAT(6X,5G5.2)
  915 FORMAT(6X,46H FOLLOWING DATA LISTED UNDER VARIABLE FMT STMT)
  916 FORMAT(  1H1)
C
C       INPUT SECTION
        NBP=0
C            DETERMINE  LIST  TYPE
    4 CONTINUE
      WRITE(6,916)
    5 READ(5,909,END=8,ERR=7) LN,LNB,FMT
      WRITE(6,908) LN,LNB,FMT
      GO TO (11,12,13,14,15,16,17,18,19),LN
      READ (5,901) (TTL(I),I=1,20)
      WRITE(6,907) (TTL(I),I=1,20)
      GO TO 8
    6 WRITE(6,912) LN
      GO TO 8
    7 WRITE(6,911)
    8 CONTINUE
      CALL EXIT
C
C        READ PROBLEM- TITLE, INITIALIZE  DATA  BASE
   11 READ(5,902) (TTL(I),I=1,15)
      DO 21 I  =16,30
   21 TTL(I)=BLANK
      DO 22 I  =1,400
   22    DB(I)=0.0
      DB(7)=5.0
      DB(8) =10.0
         DB(9) =0.001
      DB(160)=15.0
```

Figure 5. PULSE computer program (Courtesy of the Monsanto Company).

```
      DB(161)=10.
      DB(162)=-4.
      DB(163)=2.0
      DB(164)=-2.0
      DB(165)=2.0
      DB(166)=-180.
      DB(167) = 20.
      KFD=4
       DO 23 I=1,12
   23  TITLE(I)=BLANK
      GO TO 5
C         READ  CASE-TITLE
   12 READ(5,902) (TTL(I),I=16,30)
      GO TO 5
C           LIST 3- FREQUENCY RESPONSE DATA
   13 READ(5,FMT) (DB(I), I=1,5)
      GO TO 5
C          LIST 4- FREQUENCY RANGE DATA
   14 READ(5,FMT,ERR=6) (DB(I),I=6,9)
      KFD=  4
      GO TO 5
C                LISTS-FREQUENCY POINTS
   15 READ(5,FMT,ERR=6) DB(6),NF,(DB(I+100),I=1,NF)
      DB(10) = NF
      KFD=5
      GO TO 5
C            LIST6- PULSE DATA
   16 READ(5,910) (A1(I),I=1,9)
      READ(5,FMT,ERR=6) K,(PD(I,K), I=2,6),NS,(H(I,K),NBK(I,K),
     1     I=1,NS),NP,(PH(I,K),I=1,NP)
      PD(1,K)=K
      PD(7,K)=NS
      PD(8,K)=NP
      DO 24 I = 1,9
   24 PA(I,K) = A1(I)
      GO TO 5
C          PLOT CONTROL DATA
   17 READ(5,FMT,ERR=6) (DB(I),I=160,167)
   18 CONTINUE
      GO TO 5
   19 CONTINUE
      IF (LNB.EQ.0) GO TO 20
      IF (LNB.EQ.9) GO TO 30
      WRITE(6,913) LNB
      GO TO 8
C           READ DATA-ITEM CHANGES
   20 READ(5,FMT,ERR=6) J1,(I,DB(I),J=1,J1)
      GO TO 5
C
C      DATA INTERFACE SECTION
C     INPUT  DATA SUMMARY
   30 WRITE(6,903) (TTL(I),I=1,30)
      WRITE(6,904)
      IF   (DB(1).NE.0.0)
     1WRITE(6,905)(I,DB(I),I=1,5)
      WRITE(6,905) (I,DB(I),I=6,10)
      IF   (DB(5).NE.0.) WRITE(6,905) (I,DB(I),I=160,167)
C          PRINT  PULSE  DATA
      DO  31 K=1,9
      AK=FLOAT(K)
      IF (PD(1,K).NE.AK) GO TO 31
      WRITE(6,181) K, (PA(I,K),I=1,9)
      I1  =10*K+1
      I2  = I1+7
       WRITE(6,905) (I,DB(I),I=I1,I2)
      I1  =10*K+201
      NS   =PD(7,K)
      I2  =I1+NS-1
      WRITE(6,905)(I,DB(I),I=I1,I2)
C        PRINT PULSE  HEIGHT ON INITIAL ENTRY ONLY  .
      IF   (PD(5,K).EQ.0.0)   GO TO 31
      PD(5,K)=0.0
      NP  =PD(8,K)
      I1=1000*K+1
      I2  =I1+NP -1
      WRITE(6,905) (I,DB(I),I=I1,I2)
   31 CONTINUE
      WRITE(6,906)
C
      WRITE(6,906)
      NCO   =DB(2)
      NCI  = DB(1)
      KRN  =DB(3)
      KTP    =DB(4)+1.01
      KBP =DB(5)
      KFU=  DB(6)
      ND=  DB(7)
      NFD=DB(8)
      FI =DB(9)
      NF=DB(10)
      FU =RAD
      IF (KFU.EQ. 1) FU=CYC
C        SET  FREQUENCY   ARRAY  FROM  RANGE DATA
      IF   (NF.GT.0.AND. KFD.EQ.5)   GO TO 35
      ANFD=FLOAT(NFD)
      NF    =NFD*ND+1
      AN=0.0
   25 CONTINUE
      DO 32  I=  1,NFD
      FM(I)  =0.1*AINT(10.0**(1.0+AN/ANFD)+0.6)
   32      AN= AN+1.0
      DO 33   J= 1,NF
      ID    = (J-1)/NFD
      I=J- ID*NFD
   33 F(J+1) =FI*FM(I)*10.0**ID
      GO TO 38
C     SET  FREQUENCY ARRAY FROM INPUT POINTS
   35 DO  36  J=1,NF
   36 F(J+1) =DB(J+100)
C     SET ZERO FREQUENCY FOR AREA CALCULATION
   38    F(1)=0.0
      NFP1 =NF+1
      IF (NCI.GT.0) GO TO 45
C
C      COMPUTATION SECTION
C        COMPUTE TRANSFORMS ONLY
   41 K=0
```

```
   42 IF   (K.GE.9) GO TO 5
      K=K+1
      IF   (IFIX(PD(1,K)).NE.K)  GO TO 42
      GO TO 50
C        COMPUTE TRANSFORMS AND FREQUENCY
   45 K=NCI
C
C          SET PULSE PARAMETERS FOR K-TH CURVE
   50 CF =PD(4,K)
      YZ =PD(2,K)*CF
      YF=PD(3,K)*CF-YZ
      KEE=PD(6,K)
      NS= PD(7,K)
      NP=    PD(8,K)
      HM=1.0E38
      TT=0.0
      IA =1
      DO 51 I=1,NS
      HM = AMIN1(HM,H(I,K))
      NB(I)=NBK(I,K)
      TT    =TT+H(I,K)*FLOAT(NB(I)-IA)
   51 IA =NB(I)
C      ADJUST PULSE FOR TIME 0-AND CHART FACTOR
      DO 53  I=1,NP
   53      Y(I)=PH(I,K)*CF-YZ
C       ADJUST PULSE FOR LINEAR CLOSURE CORRECTION
      IF(KEE.NE.1) GO TO 60
      T=0.0
      J=1
      IB=NB(J)
      DO 56 I=2,NP
      IF(I.LE.IB) GO TO 55
      J=J+1
      IB = NB(J)
   55 T=T+H(J,K)
   56 Y(I)=Y(I)-YF*T/TT
C         COMPUTER FOURIER TRANSFORM
   60 DO 63 J=1,NFP1
      FR   =F(J)
      IF   (KFU.EQ. 1) FR=FR*6.283185
   63 CALL FTRAN (NS,NB,H(1,K),Y,YF,KEE,FR,R(J),Z(J))
C         AREA UNDER  PULSE
      AZ= SIGN(SQRT(R(1)**2 +Z(1)**2),R(1))
      IF   (NCI.GT.0)  GO TO 80
C
C         TRANSFORMS ONLY      -PRINT PULSE ID
        WRITE(6,903)       (TTL(I),I=1,30)
      WRITE(6,906)
      WRITE(6,181)K,PA(1,K),PA(2,K)
      WRITE(6,183) CF,TT,AZ
C            PRINT END CORRECTION TYPE
      IF(KEE-1) 75,72,73
   72 WRITE(6,184)
      GO TO 75
   73 WRITE(6,185)
C        PRINT TABLE HEADING AND TRANSFORM DATA
   75 WRITE(6,186)FU
      DO 77 J=2,NFP1
      A=SQRT(R(J)**2+Z(J)**2)
      P= 57.29578*ATAN2(Z(J),R(J))
      IF   (P.LT.0.0) P=P+360.0
   77 WRITE(6,187) F(J),R(J),Z(J),A,P
      GO TO 42
C
C        RESPONSE NEEDED- SAVE INPUT
   80 IF (K.EQ.NCO) GO TO 85
      CFI=CF
      HMI= HM
      KEEI=KEE
      TTI=TT
      AZI=AZ
      DO 81 J=1,NFP1
      RI(J)=R(J)
   81 ZI(J)=Z(J)
      K=NCO
      GO TO 50
C      STEADY STATE GAIN
   85 GAIN = AZ/AZI
C     FREQUENCY LIMIT BASED ON INTERVAL SIZE
      FLM= 1.570796/AMAX1(HM,HMI)
      IF(KFU.EQ.1)  FLM=FLM/6.283185
C
C      PRINT  RESPONSE TABLE
      KFL=0
      JP=0
      NPTS=0
      DO 130 J=2,NFP1
      IF(J.NE.2.AND.J.NE.38) GO TO 100
      WRITE(6,903)         (TTL(I),I=1,30)
       IF(J.NE.2) GO TO 95
C      PRINT INPUT PULSE ID
      WRITE(6,182) AIN, NCI,(PA(I,NCI),I=1,9)
      WRITE(6,183) CFI,TTI,AZI
      IF(KEEI-1)90,86,87
   86 WRITE(6,184)
      GO TO 90
   87  WRITE(6,185)
C      PRINT OUTPUT PULSE ID
   90 WRITE(6,182) OUT,K,(PA(I,K),I=1,9)
      WRITE(6,183) CF,TT,AZ
      IF(KEE-1) 93,91,92
   91 WRITE(6,184)
      GO TO 93
   92 WRITE(6,185)
   93 WRITE(6,189) FLM
      WRITE(6,190)GAIN
C    PRINT RESPONSE TABLE HEADINGS
   95 GO TO(96,97,98,99),KTP
   96 WRITE(6,191) FU
      GO TO 100
   97 WRITE(6,192) AIN,FU
      GO TO 100
   98 WRITE(6,192) OUT,FU
      GO TO 100
   99 WRITE(6,193) FU
```

Figure 5. (continued)

```
C
C         AMPLITUDE, AMP. RATIO(NORMALIZED)
  100  AI= SQRT(RI(J)**2+ZI(J)**2)
      A=  SQRT(R(J)**2+Z(J)**2)
      AR=A/AI
      IF (KRN.EQ.0.AND.MAX0(KEEI,KEE).LT.2)
     1  AR=ABS(AR/GAIN)
      ARL=ALOG10(AR)
C          PHASE ANGLES, PHASE DIFFERENCE
      PI =57.29578*ATAN2(ZI(J), RI(J))
      P =57.29578*ATAN2(Z(J), R(J))
      IF (PI.LT.0.0)  PI=PI+360.0
      IF (P  .LT.0.0)    P=P+360.0
        PL  =P-PI
C          PRINT INPUT AMPLITUDE WARNING
      IF (KFL.NE.0 .OR. AI .GE. 0.01*ABS(AZI)) GO TO 110
      WRITE(6,194)
        KFL=1
C           PRINT J TH FREQUENCY RESPONSE
 110  GO TO (111,112,113,114), KTP
 111  WRITE(6,195) F(J), AR ,  ARL, PL, RI(J), ZI(J), AI, PI,
     1              R(J), Z(J), A, P
      GO TO 120
 112  WRITE(6,195) F(J), AR, ARL, PL, RI(J), ZI(J), AI, PI
      GO TO 120
 113  WRITE(6,195) F(J),AR, ARL, PL, R(J), Z(J), A, P
      GO TO 120
 114  WRITE(6,195) F(J), AR, ARL, PL
 120  CONTINUE
C         SAVE VALUES FOR POSSIBLE BODE PLOT
 130  CONTINUE
C          END OF FREQUENCY LOOP
      GO TO 4
      END
      SUBROUTINE FTRAN(NS, NB, H, Y, YF, KEE, F, R, Z)
C     COMPUTER   =USASI FORTRAN
C     DEFINITION SECTION
C     F       FREQUENCY (RAD/TIME)
C     H       TIME BETWEEN POINTS IN JTH SECTION OF PULSE
C     IA      NUMBER OF FIRST POINT IN A SECTION
C     IB      NUMBER OF LAST POINT IN A SECTION
C     KEE     PULSE CLOSURE CODE
C     NB      NUMBER OF LAST POINT IN JTH SECTION OF PULSE
C     R       FOURIER TRANSFORM REAL PART
C     TA      TIME AT FIRST POINT IN A SECTION
C     TB      TIME AT LAST POINT IN A SECTION
C     Y       PULSE HEIGHT ARRAY
C     Z       FOURIER TRANSFORM IMAGINARY PART
C     NONCE VARIABLES-CA,CB,D,DH,FTA,FTB,FTI,IAPI,IBMI,Q,RI,SA,SB,SYI,U,
C      DECLARATIVE SECTION
      DIMENSION      NB(1), H(1),Y(1)
C      COMPUTATION SECTION
      R    =0.0
      Z   =0.0
      TA    =0.0
      IA    =1
      IF (F.EQ.0.0)    GO TO 60
C         POSITIVE FREQUENCY
      DO 50 N     =1,NS
        IB=NB(N)
C          TIME AT END OF NTH SECTION
      TB=TA+H(N)*FLOAT(IB-IA)
C        D AND Q PARAMETERS
      U=F*H(N)
      D= (SIN(U/2.0)/(U/2.0))**2
      DH=D/2.0
      IF (U.GT.0.1)   GO TO 32
      Q =U*(1.0-U**2/20.0)/6.0
      GO TO 35
 32   Q=(1.0-SIN(U)/U)/U
C       CONTRIBUTION OF END POINTS
 35   FTA=F*TA
      FTB=F*TB
      SA= SIN(FTA)
      CA= COS(FTA)
      SB=SIN(FTB)
      CB= COS(FTB)
      R=R+H(N)*(Y(IA)*(DH*CA-Q*SA)+ Y(IB)*(DH*CB+Q*SB))
      Z=Z-H(N)*(Y(IA)*(DH*SA+Q*CA)+Y(IB)*(DH*SB-Q*CB))
C               CONTRIBUTION OF INTERIOR POINTS
      IAP1    =IA+1
      IBM1    =IB-1
      IF (IAP1.GT.IBM1)  GO TO 45
      R1       =0.0
      Z1    =0.0
      DO 41 I       =IAP1,IBM1
      FTI   =F*(TA+H(N)*FLOAT(I-IA))
      R1   =R1+Y(I)*COS(FTI)
 41   Z1    =Z1+Y(I)*SIN(FTI)
      R   =R+H(N)*D*R1
      Z   =Z-H(N)*D*Z1
C          INITIALIZE FOR NEXT SECTION
 45   TA   =TB
      IA   =IB
 50   CONTINUE
C          ADJUST FOR NONCLOSURE
      IF(KEE.NE.2)  GO TO 55
      IF(YF .EQ. 0.0) GO TO 55
      R = R - YF*SIN(F*TB)/F
      Z = Z - YF*COS(F*TB)/F
   55 RETURN
C
C           ZERO FREQUENCY
   60 DO 70 N = 1,NS
      IB   =NB(N)
      SYI   =0.0
      IAP1   =IA+1
      IBM1   =IB-1
      IF (IAP1.GT.IBM1) GO TO 65
      DO 61 I   =IAP1,IBM1
 61   SYI   =SYI+Y(I)
 65   R   =R+H(N)*(0.5*(Y(IA)+Y(IB))+SYI)
 70   IA   =IB
C     ADJUST FOR NONCLOSURE
      IF (KEE .NE. 2) GO TO 80
      IF (YF .EQ. 0.0) GO TO 80
      Z= -R
      R = SIGN(1.0E-37,R)
   80 RETURN
      END
```

Figure 5. (continued)

ACKNOWLEDGMENT

A PULSE computer code included in this module largely represents a version programmed by Allen C. Pauls of the Corporate Engineering Department of the Monsanto Company, St. Louis, MO, with contributions by Stanley C. Becker. The Fourier transform algorithms were initially applied for reducing pulse test data at Monsanto in the late 1950's by Joel O. Hougen. Other major contributors to those very early data reduction efforts included Robert O. Walsh, Gerald E. Dreifke and Alan Weissner. The author has had the privilege of working closely with Dr. Hougen, Dr. Paul and others at Monsanto for several years in the application of pulse testing to determine process dynamics by experimental plant testing studies. For more specific information on the early pulse testing development at Monsanto, the following additional references are suggested:

1. Hougen, J. O., "Experiences and experiments in process dynamics," *Chem. Engr. Prog. Monograph Series*, 60, no. 4 (1964).
2. Dreifke, G. E. and J. O. Hougen, "Experimental determination of system dynamics by pulse methods," 4th JACC Conf. Proc., Minneapolis, MN, pp. 608–633, AIChE, NY (1963).
3. Hougen, J. O. and R. A. Walsh, "Pulse testing methods," Chem. Engr. Prog., 57, no. 3, pp. 69–79 (March 1961).

Module A4.2

Supervisory Control

Ram Lavie

Department of Chemical Engineering, Technion-Israel Institute of Technology Haifa, Israel

OBJECTIVES

This module aims at:

1. Familiarizing the student with the concept of supervisory process control.
2. Defining procedural steps that are necessary ingredients in the implementation of supervisory control.
3. Helping the student to appreciate some potential benefits as well as necessary efforts related to the application of supervisory control.

PREREQUISITE MATHEMATICAL SKILLS

1. Differential equations.
2. Laplace transforms.

PREREQUISITE ENGINEERING AND SCIENCE SKILLS

1. Feedback control.
2. Computer programming (Fortran).

In traditional process control, each process variable of interest is controlled separately in a feedback control loop, leaving to process design the task of minimizing interaction between the variables. Plant operating instructions must provide the operator with formulas, or tables directing the selection of set points for each control loop. Sometimes, those instructions are refined and adjusted to be functions of production objectives and environmental conditions.

The application of digital computers to process control has opened new options for an integrated approach to process control, optimization of plant operation and automation of plant supervision. Supervisory control can be defined as that part of computer process control that aims at optimizing and supervising plant operation automatically. It does not usually deal directly with the task of maintaining the process variables at their desired value by feedback control. The distinction between a supervisory control program and a "controller algorithm" is both one of function and of time scale; the supervisory control programs decides if, when and to what extent should the target value of some process variable be adjusted, while the controller algorithm relates the control action to deviations of the controlled variables from desired set points. Also, the supervisory decisions are carried out at a much slower rate than that of the control action within the control loop. Supervisory control is applicable both when the process variables are controlled in individual independent control loops or when they are controlled interdependently.

Ideally, a supervisory control program should be able to utilize information on available resources (such as feedstocks, plant conditions, etc.) together with broadly specified production objectives, to navigate the plant along a feasible, optimal, safe path automatically. It should at the same time, provide the human operator with essential reports on plant performance and problems, allowing his intervention when necessary. It will typically include procedures for alarm and emergency conditions, start-up and shut-down, safety interlocks and other on-off controls.

Supervisory control can be implemented by means of a general purpose computer, as is done in this particular exercise. However, complete process control packages (hardware and software) can be purchased from one of several suppliers on the market or may even be "tailor made" and supplied as a turn-key job by a systems house. Each avenue chosen will understandably have implications with respect to the cost, specificity of the final systems, flexibility in use, commitments concerning maintenance of both hardware and software and will depend on the availability of trained manpower in the various necessary disciplines. This exercise does not deal with all those very important aspects and is not meant to serve as an implementation guide. Since a same function may be performed differently accordingly to the system used, the student is urged to pay more attention to the ingredients of this exercise than to how exactly the functions are executed in the computer program. In different commercial systems there may be different emphasis, different organization (e.g. separation of operator performed functions from those reserved for the process engineer, such as data base generation, control configuration definition, alarm settings, etc.) and sometimes even different terminology.

The present exercise emphasizes process modeling. Firstly, it is a methodical necessity in a tutorial exercise. Secondly, it is an important, if not essential, ingredient in supervisory control. Figuratively speaking, the process model allows the utilization of the computer to extend the eyes and the brains of the operator as well as his hands. The model may be used to estimate unavailable measurements, refine existing ones and evaluate plant performance in relevant terms. It may also be used to select set points corresponding to optimal operating conditions or to some other criteria. In the present exercise, the model makes it possible to meet exactly desired production objectives through their translation into set points that are compensated for deviations due to interactions between variables and to control loop offset.

Problem Description

The process used in this exercise is a simple second–order interacting pressure system. It can be implemented as such in the lab, or simulated on a computer. It is desired to operate the process automatically, while satisfying objectives of production rate as well as some (optimal) pressure ratio. Two pressures are measured and controlled by feedback. A supervisory control program on a general purpose minicomputer (with A/D and D/A conversion capability) supervises the operation mainly by providing set points to the feedback controllers. The student is required to:

a) Analyze the process and identify problem areas, as well as the data necessary to carry out the supervisory control.
b) Calibrate the communication between process and computer.
c) Carry out one or several runs under manual and supervisory control with user supplied control subroutines.
d) Apply set point compensation for process interactions (based on a process model).

Plant Description

A flowsheet of the plant and its instrumentation is provided in Figure 1. The process itself consists of a stream of air under pressure flowing through two

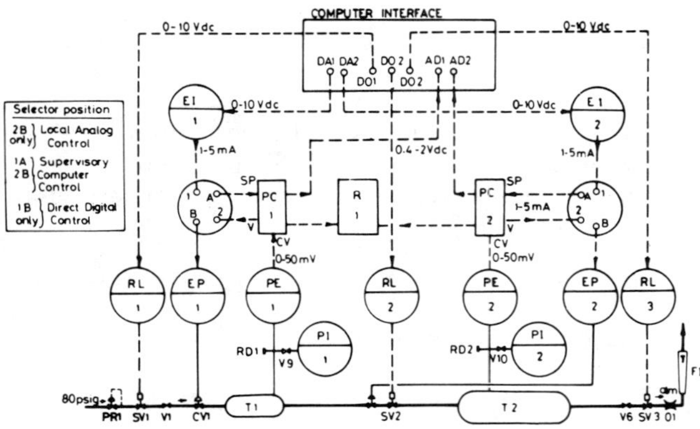

Figure 1. Process and instrumentation flowsheet.

tanks in series. The flow in and out of each tank is restricted by a control valve or a fixed restriction (orifice). The air enters the plant at 60 psig, flows through a first control valve into the first tank, then through a second control valve to the second tank and then discharged to atmospheric pressure through a restriction. The pressure (0-60 psig, 0-414 mPa) in each of the tanks is measured (PI1, PI2) and transmitted by a pressure to voltage transducer (PE1, PE2) to an analog electronic PI controller (PC1, PC2). This signal (0-50 mVdc) is available there for the generation of a feed-back control signal and it is also amplified (0-2 Vdc) and transmitted to a chart recorder (R1) as well as to one of the analog to digital converter channels (AD1, AD2) of the digital computer.

Each of the control valves (CV1, CV2) is operated by a pneumatic signal (3-15 psig, 21-104 mPa) originating in a current to pressure transducer (EP1, EP2) receiving its driving signal (1-5 mA) from one of the analog controllers (PC1, PC2) or from one of the digital to analog converter channels (DA1, DA2) of the digital computer, according to the position of a selector switch (SW1, SW2). The (0-10 Vdc) DAC computer control signal is converted to (1-5 mA) in voltage to current converters (EI1, EI2). The set points on the analog controllers (PC1, PC2) can be set either manually or from the computer according to the position of a switch on the controller.

The flow of air through the plant can be controlled by solenoid valves (S1, S2, S3) set on manually or by program originating signals (DO1, DO2, DO3) transmitted through relays (RL1, RL2, RL3). This allows for remote startup and emergency shutdown. The flow rate is indicated on rotameter (FI1).

Model

Figure 2 shows a material balance on a single tank, where:

$w \equiv$ mass flow rate

$P \equiv$ pressure

$D \equiv$ orifice size

Assuming ideal gas, an approximate model can be formulated:

$$\frac{dP_1}{dt} = \frac{RT}{VM}(w_o - w_1) = c_1'(w_o - w_1) \qquad (1)$$

with

$$w_o = \begin{cases} C_o D_o^2[(P_o - P_1)P_o]^{1/2} & (P_1 \geq 0.53\, P_o) \\ \sqrt{0.47\, c_o D_o^2 P_o} & (P_1 < 0.53\, P_o) \end{cases}$$

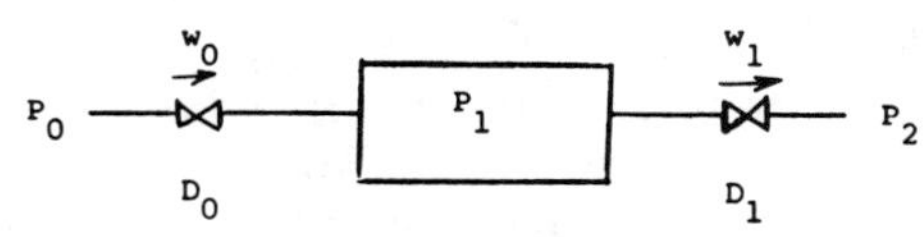

Figure 2.

$$w_1 = \begin{cases} c_1 D_1^2 [P_1 - P_2) P_1]^{1/2} & (P_2 \geqq 0.53\, P_1) \\ \sqrt{0.47}\, c_1 D_1^2 P_1 & (P_2 < 0.53\, P_1) \end{cases}$$

where the coefficients c_o, c_1, c'_1, may be functions of the temperature but do not depend to any practical extent on either the pressures P_o and P_1, or the orifice sizes D_o and D_1.

The equation illustrated by Figure 2 may be rewritten as:

$$\frac{dP_1}{dt} = k_o(D_o) \cdot \phi_o(P_1;\, P_o) - k_1(D_1) \cdot \phi_1(P_1;\, P_2)$$

$$= f_1(P_o,\, P_1,\, P_2;\, D_o,\, D_1) \qquad (1a)$$

with

$$k_o(D_o) = \left(\frac{RTc_o}{VM}\right) D_0^2;\; k_1(D_1) = \left(\frac{RTc_1}{VM}\right) D_1^2$$

and

$$\phi_o(P_1;\, P_o) = \begin{cases} [(P_o - P_1)P_o]^{1/2} & \text{where } P_1 \geqq 0.53\, P_o \\ \sqrt{0.47}\, P_o & \text{where } P_1 < 0.53\, P_o \end{cases}$$

$$\phi_1(P_1;\, P_2) = \begin{cases} [(P_1 - P_2)P_1]^{1/2} & \text{where } P_2 \geqq 0.53\, P_1 \\ \sqrt{0.47}\, P_1 & \text{where } P_2 < 0.53\, P_1 \end{cases}$$

The same model applies to the second pressure as well:

$$\frac{dP_i}{dt} = f_i(P_{i-1},\, P_i,\, P_{i+1},\, D_{i-1},\, D_i), \quad i = 1, 2 \qquad (2)$$

Linear Model

The process is clearly nonlinear. Nevertheless, it can be controlled reasonably well on the basis of a linearized model.

Depending on the range of expected operating pressures, one may use a local linear approximation (Taylor series expansion) around an operating point, or an average linear approximation over a wide range of operating pressures.

The linearized model will have the form

$$\frac{dP_1}{dt} = a_{11} P_1 + a_{12} P_2 + b_{11} D_o + b_{12} D_1 + b_{14} P_o$$

$$\frac{dP_2}{dt} = a_{21} P_1 + a_{22} P_2 + b_{22} D_1 + b_{23} D_2 + b_{25} P_3 \qquad (3)$$

described in the block diagram (Figure 3).

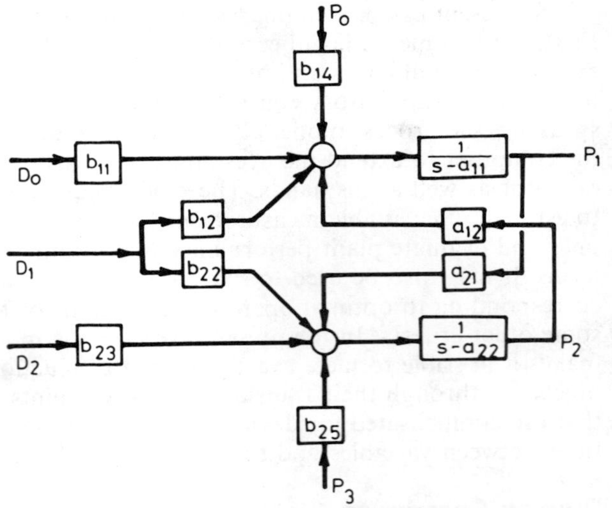

Figure 3. Linearized plant model block diagram.

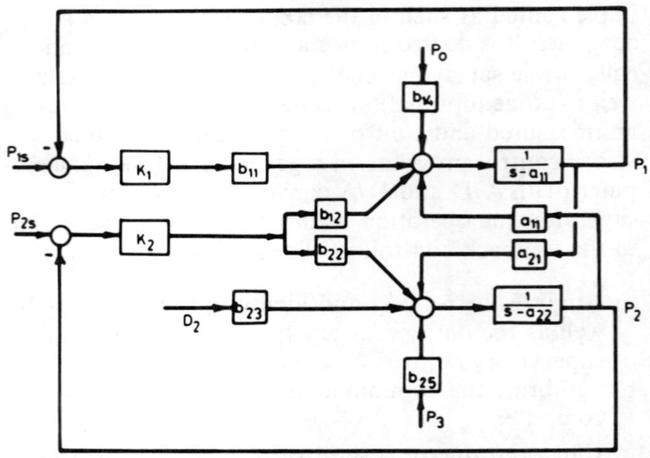

Figure 4. Controlled process block diagram.

Set Point Compensation for Interactions and Offset

As evident in Equations 3, the pressures P_1 and P_2 are interacting with each other, so that any control action that is applied in order to influence one pressure, will inevitably influence the other. Consider the process in Figure 1, with the flow into tank 1 controlled by CV1 in a feedback proportional control loop according to P_1, while the flow into tank 2 is similarly controlled by CV2 according to P_2.

Figure 4 is a block diagram of the controlled process. The supervisory control program should keep the process at desired states P_{1ss} and P_{2ss} by set points P_{1s} and P_{2a}. However, the pressures P_1, P_2 respond to changes in set points according to:

$$P_1 = \frac{1}{s - (a_{11} - K_1 b_{11})} [(a_{12} - K_2 b_{12}) P_2 + K_1 b_{11} P_{1s} + K_2 b_{12} P_{2s} + L_1]$$

$$P_2 = \frac{1}{s - (a_{22} - K_2 b_{22})} [a_{21} P_1 + K_2 b_{22} P_{2s} + L_2] \qquad (4)$$

where L_1 and L_2 are loads that are functions of P_o, P_3, D_2 which are constants in this problem.

For constant loads (servo system), the steady-state response is described by:

$$(K_1 b_{11} - a_{11}) p_{1ss} + (K_2 b_{12} - a_{12}) p_{2ss} = K_1 b_{11} p_{1s} + K_2 b_{12} p_{2s} + L_1$$

$$-a_{21} p_{1ss} + (K_2 b_{22} - a_{22}) p_{2ss} = K_2 b_{22} p_{2s} + L_2 \qquad (5)$$

Let

$$Q=\begin{bmatrix}(K_1 b_{11}-a_{11}) & (K_2 b_{12}-a_{12})\\ -a_{21} & (K_2 b_{22}-a_{22})\end{bmatrix}$$

$$R=\begin{bmatrix}k_1 b_{11} & K_2 b_{12}\\ k_2 b_{22} & 0\end{bmatrix}$$

$$\boldsymbol{P}_{ss}=\begin{bmatrix}P_{1ss}\\ P_{2ss}\end{bmatrix}$$

$$\boldsymbol{P}_s=\begin{bmatrix}P_{1s}\\ P_{2s}\end{bmatrix};$$

$$L=\begin{bmatrix}L_1\\ L_2\end{bmatrix}$$

then,

$$\boldsymbol{P}_s=R^{-1}\{\boldsymbol{P}_{ss}-\boldsymbol{L}\} \tag{5a}$$

Given desired p_{1ss} and p_{2ss}, then p_{1s} and p_{2s} can be evaluated by solving the two linear algebraic equations in two unknowns. Thus the supervisory control program can specify computed set points p_{1s} and p_{2s} to the feedback controllers. *Note*: Set point compensation is used here as a sample application of supervisory control, even though it is not practiced very often because of its dependence on a model.

Implementation

The computer program featured in the appendix to this module was prepared specifically for this exercise. It includes most of the functions typical to supervisory control. It may be run as it is (provided of course the computer I/O channels are connected to the process described in the plant description section, and thus provide elementary supervisory control. All necessary calibration parameters are communicated to the program in an initial dialogue. Those parameters can be evaluated through a calibration routine described in the next section. A more accurate control of the process can be achieved by the same program, provided the standard (dummy) compensation subprogram is replaced by one that performs this function more effectively, using Equation 5. However, this equation uses process parameters that can be obtained through a calibration experiment as detailed shortly. Those parameters must be included explicitly in the subroutine prepared by the student.

A. *Calibration*

1. Sensor Calibration.
   a) By means of the I (input) command, bring the process to various steady–state pressures p_1, p_2 and flow rates w_2 over the range of interest.
   b) Plot the pressure read on the pressure gages versus the voltage signals transmitted from the process to the computer as indicated on the terminal in response to a L command.
   c) Plot the flow rate indicated on the rotameter versus the pressure p_2.
   d) Evaluate the parameters a_1, b_1, a_2, b_2 in:

$$\underset{\text{computer}_1}{p\,(\text{volts})}=a_1\,\underset{\text{gage}_1}{p\,(\text{psig})}+b_1$$

$$\underset{\text{computer}_2}{p\,(\text{volts})}=a_2\,\underset{\text{gage}_2}{p\,(\text{psig})}+b_2 \tag{6}$$

   e) According to Equation 2, the flow rate w_2 is a function of p_2, possibly of p_3 (a constant) and a single parameter:

$$a_3=k_2=C_2 D_2^2$$

   Evaluate this parameter. Note that this is best done in the critical flow range.

2. Evaluation of process parameters.
   a) By means of the I command, transmit a pair of set points p_{1s}, p_{2s} (in the vicinity of the desired operating range: $p_1=50$ psig, 345 mPa, $p_2=30$ psig, 207 mPa) to the controllers. Wait for steady state, and record resulting pressures p_{1ss}, p_{2ss}. Repeat for different pairs of set points.
   b) Use the results to evaluate the parameters relating the resulting steady state values to the setpoints as in Equation 5a.

   Note that since the controller gains K_1, K_2 are included in the evaluated parameters, any further manipulation of these gains will require renewed evaluation of the process parameters.

B. *Program Preparation*

Several standard tasks to be performed by the supervisory control program such as alarm and emergency conditions monitoring and data logging have been programmed within the available program. However, the main control function selecting set points for supervisory control (or control valve position in direct digital control) has been left for the student to program as a Fortran subroutine to the main program (see Appendix A).

Basically, this subroutine accepts two measurements, P1 and P2, together with production objectives and calibration data communicated through COMMON and then generates two commands, DV1 and DV2, sent through D/A conversion to the analog controller remote set points (or directly to the controller valves in the case of direct digital control). Format convention and sample subroutine are brought as an appendix to the computer program. This user–supplied subroutine (which includes, for example, optimization procedures or supervisory compensation considerations) should be prepared, debugged and added to the main program beforehand.

1. A master computer control program in the appendix to this module. Study carefully its operating instructions and appreciate the options available in its operation in terms of operator intervention to adjust process and control parameters, operating conditions and production objectives, control rates, alarm setting, data logging, plotting of history, etc. Make sure you understand

how your own control subprogram fits in. Note the names given to the variables that are communicated through COMMON between your subprogram and the master program.

2. Write your own control subprograms for supervisory control that:
   a) Translates production objectives into set point directives to the process. The production objectives are formulated in terms of

      1. Desired production rate PRODR(scfh)=
      2. Desired pressure distribution in the process, coded as

         1≡minimize (P_1/P_2)

         2≡maximize (P_1/P_2)

         3≡constant ratio (P_1/P_2); PRATIO$=(P_1/P_2)$

   b) If necessary, specifies set points as function of time.
   c) Adjusts set points to compensate for interactions.
3. Debug and test your programs independently from the master control program or the process.

C. *Running the Control Program*

a) Build a task incorporating the master control program and your own supervisory control subprogram according to instructions.
b) Set manual valves on the process as in Figure 1.
c) Check wiring connections (and proper grounding) between process - computer interface - computer.
d) Set selector switches to supervisory control position.
e) Run control task.
   (Each run is limited in time by the amount of data that can be accumulated in storage buffer for display as plots at end of run).
   Run 1 - (general program testing)
   Analog control algorithm: proportional
   Log every 20 seconds
   Control every 1 second
   Production rate - try different values.
   Pressure distribution - try different options.
   Observe process instruments, log records.
   Plot history for P(1), P(2), DV(1), DV(2).
   Run 2 - (Supervisory with/without compensation)
   Same as Run 1 except
   Production rate - as necessary for P(2)=30 psig(207 mPa)
   Pressure distribution - P(1)=50 psig(345 mPa)
   When pressures reach a steady state, activate compensation algorithm and observe response.

ACKNOWLEDGMENT

This module was originally developed at the Technion, reworked and tested at the University of Wisconsin where the author visited during 1976/77, and then again revised and tested in a somewhat different form at the Technion. The author is indebted to all those (both at UW and the Technion) who contributed in various forms to the evolution of the problem. Special thanks are due to Dennis Arnon who wrote the computer program and its operating instructions.

LITERATURE CITED

1. W. L. Luyben, "Process Modelling, Simulation and Control for Chemical Engineers," Chapters 11, 15, McGraw-Hill (1973).
2. Smith, Cecil L., "Digital Computer Process Control," Chapters 1, 3, 8, Intext (1972).
3. Perry, R. H. & C. H. Chilton, "Chemical Engineers Handbook," Chapter 22, 5th Ed. McGraw-Hill (1973).

STUDY PROBLEMS

1. Linearize the process model Equations 2 to yield Equation 3. Express explicitly every parameter a_{ij}, b_{ij}, in terms of the original parameters in Equation 2.

HOMEWORK PROBLEMS

1. Show that the matrix R in Equation 5a can always be inverted (provided $K_2 \neq 0$).
2. Discuss the effects of set point compensation in run 2.
3. Can the compensation algorithm used in run 2 be applied as it is to different operating conditions?
4. The control loop in the program is activated at spaced intervals (some processor time must be allowed for performing other functions). Discuss the effect of this interval on the quality of supervisory control.
5. Is a process model always essential to the implementation of supervisory control? Discuss.
6. Can set point compensation deal effectively with changes in load? Discuss.
7. Supervisory control is sometimes used to navigate a process along a desired path in time. Can you see any limitation on the rate of change in such a path? Discuss.

APPENDIX

Program Name: *PLC*
Class: *Normal*
Type: *Self-Standing + Subroutine*

Title: Plant Control Program
Author: Dennis Arnon
University of Wisconsin

Abstract:
The program (Figure 5) provides the necessary environment for computer control of a process by means of a user-defined algorithm. It consists of three tasks, run simultaneously at different priorities:

1. The Master task provides the communication between the user and the program through an interactive dialogue.

Program Listing.

```
C MASTER-----
C AP 17 77 DSA
        COMMON /FLAGS/ IFLG(2),NUMFLG
        COMMON /CHAN/ ICHAD(2),ICHDA(5),NCHAD,NCHDA
        COMMON /DATEL/ AD(2),DA(5)
        COMMON /LUN/ LUNAD.LUNTI.LUNPLT,LUNDMP
        COMMON /CALIB/ A(3),B(3),PUNITS,FUNITS.NUMCAL
        COMMON /BOUNDS/ PHALM(2),PLALM(2),PEMG(2),NUMALM
        COMMON /GOALS/ PRODR,PRODO,PRATIO
        COMMON /MODE/ CURMOD,DIRDIG,SUPERV
        COMMON /TIMINT/ CNTDLY,LOGDLY,CNTUN
        COMMON /ACTUAL/ P(2),F
        COMMON /CNTRL/ DV(2)
        COMMON /OBJCTV/ SET(2)
        COMMON /CNTLC/ ICOUNT,CK(2),TI(2).TD(2)
        COMMON /SPACE/ ITOP,IMAX,LREC(220,6),RREC(220,7)
        REAL*8 PUNITS,FUNITS
        INTEGER PRODO
        LOGICAL*1 CURMOD,DIRDIG,SUPERV
        INTEGER CNTDLY
        LOGICAL*1 CNTUN
        LOGICAL*1 LREC
C
C       ... TASK NAMES IN RADIX 50
        DATA CNTTNM/6RCNTMAN/
        DATA RECTNH/6RRECCDR/
        REAL LOGTNM
        DATA LOGTNM/6RLOGMAN/
C
C       ... DUMMY ARRAY FOR DIRECTIVES AND DTOAC
        DIMENSION IDUM(4)
C       ... DIRECTIVE TIME UNITS
        DATA ITICK,ISEC,IMIN,IHOUR/1,2,3,4/
C
C       ...COMMAND LETTERS
        LOGICAL*1 COM(14),CM
        DATA COM/'H','C','B','G','M','X','Y','L','D','E','I','R','P','S'/
        DATA NUMCOM/14/
C
C       ...FOR I COMMAND
        LOGICAL COND
C
C       ...FOR P COMMAND
        DIMENSION IPLT(3)
C
C       ...INITIALIZE HISTORY FILE POINTER
        ITOP = 0
C
C       ...CLEAR ALL FLAGS
        DO 10 I = 1,NUMFLG
 10        IFLG(I) = 0
C
C       ...LOGICAL UNITS FOR HISTORY FILE DUMP AND PLOTTING
        CALL ASSIGN(LUNDMP, 'LP:')
C       ......ALWAYS USED SAME FILE FOR PLOTTING UMF,WRITE
        CALL ASSIGN( LUNPLT, 'DK1:FOR003.DAT;1')
C
C       ...INITIALIZE NO. OF EXECTNS OF CNTL SUBR
        ICOUNT = 0
C
C       ...INITIALIZE VALUES TO 0; SHUT OFF PLANT
        F = 0.
        DO 15 I = 1,2
           P(I) = 0.
 15        SET(I) = 0.
        DO 20 I = 1,NCHDA
 20        CALL SETDA(I, 0.)
C
C       ...COUNT NUMBER OF EXECUTIONS
        IEXCNT = 0
C
C       ...NUMBER OF PRESCRIBED INITIAL COMMANDS
        DATA NUMIN /7/
C
C ........INTERPRETER LOOP .......
 100    IF (IEXCNT .LE. NUMIN) IEXCNT = IEXCNT + 1
        IF (IEXCNT .LE. NUMIN) GO TO(200,300,400,500,600,700,800),IEXCNT
        TYPE 9000
 9000   FORMAT(' $$')
C
C       ... CHECK FOR STORAGE OVERFLOW
        IF (IFLG(2) .EQ. 0) GO TO 110
           TYPE 9005,ITOP
 9005   FORMAT(' ***STORAGE OVERFLOW,ITOP=',15)
 110    ACCEPT 9010,CM
 9010   FORMAT(A1)
        DO 120 I = 1,NUMCOM
 120       IF (CM .EQ. COM(I) ) GO TO 140
        TYPE 9020,CM
 9020   FORMAT('  ',A1,' IS NOT A LEGAL COMMAND')
        GO TO 100
 140    GO TO (200,300,400,500,600,700,800,900,1000,1100,1200),I
        GO TO (1300,1400,1500),I - 11
C
C       ... H COMMAND (CHANNEL ASSIGNMENTS)
 200    DO 210 I = 1,NCHDA
           TYPE 9030,I
 9030      FORMAT(' ICHDA(',I1,')?  16')
           ACCEPT 9040, ICHDA(I)
 9040      FORMAT(I6)
 210       CONTINUE
        DO 220 I = 1,NCHAD
           TYPE 9050,I
 9050      FORMAT(' ICHAD(',I1,')?  16')
           ACCEPT 9060,ICHAD(I)
 9060      FORMAT(I6)
 220       CONTINUE
        GO TO 100
C
C       ... C COMMAND (CALIBRATION CONSTANTS)
 300    DO 310 I = 1,NUMCAL
           TYPE 9070,I
 9070      FORMAT(' A',I1,'? F10.4')
           ACCEPT 9080,A(I)
 9080      FORMAT(F10.4)
C
C          ...B(3) NO LONGER USED AP 27 77
           IF (I .EQ. 3) GO TO 310
           TYPE 9090,I
 9090      FORMAT(' B',I1,'? F10.4')
           ACCEPT 9080,B(I)
 310       CONTINUE
        TYPE 9100
 9100   FORMAT(' P UNITS?  AB')
        ACCEPT 9110,PUNITS
 9110   FORMAT(AB)
        Type 9120
 9120   FORMAT(' F UNITS?  A8')
        ACCEPT 9110,FUNITS
        GO TO 100
C
C       ... B COMMAND (ALARM AND EMERGENCY BOUNDS)
 400    DO 410 I = 1,NUMALM
           TYPE 9130,I
 9130      FORMAT(' PHALM(',I1,')?Flo.4')
           ACCEPT 9140,PHALM(I)
 9140      FORMAT(F10.4)
 410       CONTINUE
        GO TO 100
C
C       ... G COMMAND ( PRODUCTION GOALS)
 500    TYPE 9150,FUNITS
 9150   FORMAT(' PRODR?(',AB,')  F10.4')
        ACCEPT 9160,PRODR
 9160   FORMAT(F10.4)
        TYPE 9170
 9170   FORMAT(' PRODO?(1=MINPRES,2=MAXPRES,3=CONSTPR)')
        ACCEPT 9180,PRODO
 9180   FORMAT(I6)
        If (PRODO.LT.3) GO TO 100
        TYPE 9190
 9190   FORMAT(' PRATIO?  F10.4')
        ACCEPT 9160,PRATIO
        GO TO 100
C
C       ... M COMMAND (MODE - DIRECT DIGITAL OR SUPERVISSORY)
 600    TYPE 9200
 9200   FORMAT(' CURMOD? (1=DIRECT DIGITAL,2=SUPERVISORY)')
        ACCEPT 9210,CURMOD
 9210   FORMAT(I6)
        GO TO 100
C
C       ...X COMMAND (SET DELAY BETWEEN LOGS)
 200    TYPE 9220
 9200   FORMAT(' LOGDLY? (SECONDS)  I6')
        ACCEPT 9230,LOGDLY
 9230   FORMAT(I6)
C
C       ... CANCEL PREVIOUS DELAY
        CALL CANALL( LOGTNM, IDS)
        IF (IDS .NE. 1) TYPE 9240,IDS
 9240   FORMAT(' IDS=',I6,' T COMMAND, LOGMAN CANALL')
C
C       ... DO NOT REINITIATE IF DELAY < 0
        IF (LOGDLY .LT. 0) GO TO 100
        CALL RUN( LOGTNM, IDUM, 1, ITICK, LOGDLY, ISEC, IDS)
        IF (IDS .NE. 1) TYPE 9250,IDS
 9250   FORMAT(' IDS=',I6,' T COMMAND, LOGMAN RUN')
        GO TO 100
C
C       ... Y COMMAND (SET DELAY BETWEEN CONTROL DECISIONS)
 800    TYPE 9260
 9260   FORMAT(' CNTUN (1=TICKS,2=SECONDS)  I6')
        ACCEPT 9270,CNTUN
 9270   FORMAT(I6)
        TYPE 9280
 9280   FORMAT(' CNTDLY?  I6')
        ACCEPT 9270,CNTDLY
C
C       ... CANCEL PREVIOUS DELAY
        CALL CANALL( CNTTNM, IDS)
        IF (IDS .NE. 1) TYPE 9290,IDS
 9290   FORMAT(' IDS=',I6,' T COMMAND,CNTMAN CANALL')
C
C       ... DON'T REINITIATE, DON'T OPEN CONNECTING VALUES IF DELAY < 0
        IF (CNTDLY .LT. 0) GO TO 100
        CALL RUN(CNTTNM, IDUM, 1, ITICK, CNTDLY, CNTUN, IDS)
```

Figure 5. PLC—The plant control program listing.

```
      IF (IDS .NE. 1) TYPE 9300,IDS
 9300 FORMAT(' IDS =', I6,' T COMMAND, CNTMAN RUN')
C
C     ...... OPEN VALVE ALLOWING AIR TO BE RECEIVED FROM SOURCE
      CALL SETDA(3, 10.)
C
C     ...... OPEN VALVE BETWEEN THE TANKS
      CALL SETDA(4, 10.)
      GO TO 100
C
C     ... L COMMAND (SPECIAL LOG)
 900  CALL REQUES(LOGTNM, IDUM, IDS)
      IF (IDS .NE. 1) TYPE 9310,IDS
 9310 FORMAT(' IDS=',I6,' L COMMAND')
      GO TO 100
C
C     ... D COMMAND (DUMP HISTORY FILE ON LINE PRINTER)
 1000 DO 1010 I = 1,ITOP
         WRITE(LUNDMP, 9350) (LREC(I,J),J=1,5),(RREC(I,J),J=1,6),
     X     LREC(I,6),RREC(I.7)
 9350    FORMAT(1X,3(I2,1X,I1,6(1X,F10.4),1X,I1,1X,F10.4)
 1010    CONTINUE
      GO TO 100
C
C     ... E COMMAND (SHUT OFF PLANT, TERMINATE SUBTASKS, EXIT)
 1100 DO 1110 I = 1,NCHDA
LLL)     CALL SETDA(I, 0.)
      TYPE 9410
 9410 FORMAT(' MASTER AND SUBTASKS ARE TERMINATING...')
      CALL CANALL( CNTTNM, IDS)
      IF (IDS .NE. 1) TYPE 9420,IDS
 9420 FORMAT(' IDS=',I6,' E COMMAND,CNTMAN CANCEL')
      CALL CANALL(LOGTNM,IDS)
      IF (IDS .NE. 1) TYPE 9430,IDS
 9430 FORMAT(' IDS=',I6,' E COMMAND,LOGMAN CANCEL')
      CALL EXIT
C
C     ... I COMMAND (IMMEDIATE VOLTAGE TO D/A CHANNEL)
 1200 TYPE 9360
 9360 FORMAT(' CHANNEL?  I6')
      ACCEPT 9370,ICHAN
 9370 FORMAT(I6)
      TYPE 9380
 9380 FORMAT(' VOLTAGE? (IN RANGE 0. TO 10.)  F10.4')
      ACCEPT 9390,VOLT
 9390 FORMAT(F10.4)
C
C     ...CHECK FOR REQUESTED VOLTAGE OUT OF RANGE
      COND = ( VOLT .LT. 0.). OR. (VOLT .GT. 10.)
      IF (COND) TYPE 9400
      IF (COND) GO TO 100
 9400 FORMAT(' REQUEST REJECTED')
      IDUM(1) = INT(VOLT * 204.7)
      CALL DTOAC( ICHAN, ICHAN, IDUM, LUNAD)
      GO TO 100
C
C     ... R COMMAND (RESET TOP OF STORAGE AND FLAGS)
 1300 ITOP = 0
C     ... RESET NUMBER OF CALLS TO CNTL
      ICOUNT = 0
      DO 1310 I = 1,NUMFLG
 1310    IFLG(I) = 0
      GO TO 100
C
C     ... F COMMAND (PLOT UP TO THREE VARIABLES)
 1400 IF (CNTDLY .GE. 0) TYPE 9470
      IF (CNTDLY .GE. 0) GO TO 100
 9470 FORMAT(' PLOTTING ILLEGAL WHEN CONTROL SUBTASK RUNNING')
      TYPE 9440)
 9440 FORMAT(' NO. OF VARIABLES TO BE PLOTTED? (1,2, OR 3)')
      ACCEPT 9450,NPLT
 9450 FORMAT(I6)
      IF (NPLT .LT. 0 .OR. NPLT. GT. 3) GOT TO 1400
      DO 1410 I = 1,NPLT
         TYPE 9460
 9460    FORMAT(' NEXT VAR? (1=P1,2=P2,3=SET1,4=SET2,5=F)')
 1410    ACCEPT 9450,IPLT(I)
      REWIND LUNPLT
      DO 1420 I = 1,ITOP
 1420    WRITE(LUNPLT) (RREC(I,IPLT(J)), J=1,NPLT), FLOAT(I)
      CALL PLOT3( ITOP, NPLT, LUNPLT)
      GO TO 100
C
C     ... S COMMAND ( PARAMETERS FOR CNTL SUBR)
 1500 DO 1510 I = 1,2
         TYPE 9480,I
 9480    FORMAT(' CK(',I1,')?  F10.4')
         ACCEPT 9490,CK(I)
 9490    FORMAT(F10.4)
         TYPE 9500,I
 9500    FORMAT(' TI(',I1,'(?  F10.4')
         ACCEPT 9490,TI(I)
         TYPE 9510,I
 9510    FORMAT(' TD(',I1,')?  F10.4')
         ACCEPT 9490,TD(I)
 1510    CONTINUE
      GO TO 100
      END

      SUBROUTINE SETDA(I,V)
```

```
C  INITIATE D?A CONVERSION ON THE ITH D?A CHANNEL (AS DEFINED BY
C       THE USER WITH THE H COMMAND IN MASTER) AND SET DA(I) = V.
C AP 17  77 DSA
      COMMON /FLAGS/ IFLG(2),NUMFLG
      COMMON /CHAN/ ICHAD(2),ICHDA(5),NCHAD,NCHDA
      COMMON /DATEL/ AD(2),DA(5)
      COMMON /LUN/ LUNAD,LUNTI,LUNPLT,LUNDMP
      COMMON /CALIB/ A(3),B(3),PUNITS,FUNITS,NUMCAL
      COMMON /ROUNDS/ PHALM(2),PLALM(2),PEMG(2),NUMALM
      COMMON /COALS/ PRODR,PRODO,PRATIO
      COMMON /MODE/ CURMOD,DIRDIG,SUPERV
      COMMON /TIMINT/ CNTDLY,LOGDLY, CNTUN
      COMMON /ACTUAL/ P(2),F
      COMMON /CNTRL/ DV(2)
      COMMON /OBJCTV/ SET(2)
      COMMON /CNTLC/ ICOUNT,XK(2),TI(2),TD(2)
      COMMON /SPACE/ ITOP,IMAX,LREC(220,6),RREC(220,7)
      REAL*8 PUNITS,FUNITS
      INTEGER PRODO
      LOGICAL*1 CURMOD,DIRDIG,SUPERV
      INTEGER CNTDLY
      LOGICAL*1 CNTUN
      LOGICAL*1 LREC
C
C     ... DUMMY ARRAY FOR DTOAC
      DIMENSION IDUM(1)
      LOGICAL COND
C
C     ... CHECK FOR OUT OF RANGE VOLTAGE
      COND = ( (V .LT. 0.) .OR. ( V .GT. 10.))
      IF (COND) TYPE 1000,V
      IF (COND) RETURN
 1000 FORMAT(' *** SETDA - ILLEGAL VOLTAGE OF ',F10.4,'VOLTS '
     X ' REQUEST IGNORED')
      DA(I) = V
      IDUM(1) = INT( V * 204.7)
      CALL DTOAC(ICHDA(I), ICHDA(I), IDUM, LUNAD)
      RETURN
      END
C     LOGMAN---
C AP 17  77 DSA
      COMMON /FLAGS/ IFLG(2),NUMFLG
      COMMON /CHAN/ ICHAD(2),ICHADA(5),NCHAD,NCHDA
      COMMON /DATEL/ AD(2),DA(5)
      COMMON /LUN/ LUNAD,LUNTI,LUNPLT,LUNDMP
      COMMON /CALIB/ A(3),PUNITS,FUNITS,NUMCAL
      COMMON /SOUNDS/ PHALM(2),PLALM(2),PEMG(2),NUMALM
      COMMON /GOALS/ PRODR,PRODO,PRATIO
      COMMON /MODE/ CURMOD,DIRDIG,SUPERV
      COMMON /TIMINT/ CNTDLY,LOGDLY,CNTUN
      COMMON /ACTUAL/ P(2),F
      COMMON /CNTRL/ DV(2)
      COMMON /OBJCTV/ SET(2)
      COMMON /CNTLC/ ICOUNT,CK(2),TI(2),TD(2)
      COMMON /SPACE/ ITOP,IMAX,LREC(220,6),RREC(220,7)
      REAL*S PUNITS,FUNITS
      INTEGER PRODO
      LOGICAL*1 CURMOD,DIRDIG,SUPERV
      INTEGER CNTDLY
      LOGICAL*1 CNTUN
      LOGICAL*1 LREC
      REAL&8 POBJ(3),MODEC(2)
      DATA POBJ/'MINPRES ','MAXPRES ','CONSTPR '/
      DATA MODEC/'DIR DGTL',"SUPERVIS'/
      DATA ICONPR/3/
      CALL TIME(IH,IM,IS,IT)
      WRITE(6,100) IH,IM,IS,IT,POBJ(PRODO),MODEC(CURMOD),
     X  (I,P(I),I,SET(I),I=1,2),F,PRODR
 1000 FORMAT('0',3(I2,':'),I2,' ',
     X  A8,' ',AB,/,2(' P(',I1,')=',F10.4,' SET(',I1,')=',
     X F10.4),/,'       FA=',F10.4,' FR=',F10.4)
      IF (PRODO .EQ. ICONPR) WRITE(6,1010) PRATIO
 1010 FORMAT('        REQUESTED CONSTANT PRESSURE RATIO=',F10.4)
      CALL EXIT
      END
C--CNTMAN --
C AP 17  77 DSA
      COMMON /FLAGS/ IFLG(2),NUMFLG
      COMMON /CHAN/ ICHAD(2),ICHDA(5),NCHAD,NCHDA
      COMMON /DATEL/ AD(2),DA(5)
      COMMON /LUN/ LUNAD,LUNTI,LUNPLT,LUNDMP
      COMMON /CALIB/ A(3),B(3),PUNITS,FUNITS,NUMCAL
      COMMON /BOUNDS/ PHALM(2),PLALM(2),PEMG(2),NUMALM
      COMMON /GOALS/ PRODR,PRODO,PRATIO
      COMMON /MODE/ CURMOD,DIRDIG,SUPERV
      COMMON /TIMINT/ CNTDLY,LOGDLY,CNTUN
      COMMON /ACTUAL/ P(2),F
      COMMON /CNTRL/ DV(2)
      COMMON /OBJCTV/ SET(2}
      COMMON /CNTLC/ ICOUNT,CK(2),TI(2),TD(20
      COMMON /SPACE/ ITOP,IMAX,LREC(220,6),RREC(220,7)
      REAL*8 PUNITS,FUNITS
      INTEGER PRODO
      LOGICAL*1 CURMOD,DIRDIG,SUPERV
      INTEGER CNTDLY
      LOGICAL*1 CNTUN
      LOGICAL*1 LREC
C
C     ... DUMMY ARRAY FOR ATODC
      DIMENSION IDUM(1)
```

Figure 5. (continued)

```
C
C         ... AP 19 77 PROCEED FLAG (IFLG(1)) NO LONGER USED
C
C         ... COMPUTE NEW P(1),P(2)
          DO 10 I = 1,2
             CALL ATODC( ICHAD(I),ICHAD(I), IDUM, LUNAD)
             AD(I) = FLOAT( IDUM(1) ) / 204.7
 10          P(I) = ( AD(I) - B(I) ) / A(I)
C
C         ... CHECK FOR EXCEEDING ALARM
          IF ( P(1) .LT. PHALM(1) .AND. P(2) .LT. PHALM(2) ) GO TO 30
D         CALL BELL(LUNTI)
          CALL TIME(IH,IM,IS,IT)
D         WRITE(LUNTI,1000) IH,IM,IS,IT
 1000     FORMAT(' *** ',3(I2,';'),I2,' ALARM CONDITION IN PLANT')
          CALL BELL(6)
          WRITE(6,1000) IH,IM,IS,IT
C
C         ... COMPUTE ACTUAL FLOW RATE
 30       F = SORT(0.47) * A(3) * (P(2) + 14.7)
          IF (P(2),LT.13.) F = A(3) * (P(2)**2 + 14.7 * P(2))
C
C         ... COMPUTE NEXT CONTROL DECISION
          CALL CNTL
C
C         ... CHECK FOR OUT OF RANGE VALUE COMPUTED BY CNTL
          DO 40 I = 1,2
             IF ( DV(I) .LT. 0. .OR. DV(I) .GT. 1.) GO TO 50
 40          CONTINUE
          GO TO 60
D50        CALL BELL(LUNTI)
C ---     CONTINUE INSERTED 4/20/77 SINCE CALL SELL9LUNTI) EXCLUDED
 50        CONTINUE
D          WRITE(LUNTI,1010) DV(1),DV(2)
           CALL BELL(6)
           WRITE(6,1010) DV(1),DV(2)
 1010      FORMAT(' *** CONTROL COMMAND OUT OF RANGE; DV(1),DV(2) =',
    X       F10.4,',',F10.4)
           CALL TIME(IH,IM,IS,IT)
           WRITE(6,1000) IH,IM,IS,IT
C
C          ... SEND CONTROL COMMAND TO DATEL - JUST NEED TO MULTIPLY BY 10
 60        DO 70 I = 1,2
 70           CALL SETDA(I, 10. * DV(I) )
C
C          ... WRITE SAMPLE RECORD
C          ...... CHECK FOR HISTORY FILE OVERFLOW; SET FLAG IF SO
 75        IF (ITOP .EQ. IMAX) IFLO(2) = 1
           IF (ITOP .EQ. IMAX) CALL EXIT
           ITOP = ITOP + 1
           I = ITOP
           CALL TIME(IH,IM,IS,IT)
           LREC(I,1) = IH
           LREC(I,2) = IM
           LREC(I,3) = IS
           LREC(I,4) = IT
           LREC(I,5) = CURMOD
           LREC(I,6) = PRODO
           DO 80 J = 1,2
              RREC(I,J) = P(J)
C
C          ... SET IS PERCENTAGE IN DIRDIG,PRESSURE IN SUPERV
           IF (CURMOD.EQ.DIRDIG) SET(J) = 10, * DA(J)
           IF (CURMOD,EQ.SUPERV) SET(J) = (AD(J) - B(J)) / A(J)
 80        RREC(I,2+J) = SET(J)
          RREC(I,5) = F
          RREC(I,6) = PRODR
          RREC(I,7) = PRATIO
          CALL EXIT
          END
```

Figure 5. (continued)

2. The CNTMAN task is the one that actually monitors the process, communicating with it through the D/A and A/D channels. Its course of action, following a user-defined CNTL subroutine, is controlled through parameters that are COMMON to the CNTMAN and MASTER tasks. It is activated periodically as requested by the user and it has the highest priority. It also keep records in a buffer.
3. The LOGMAN task provides the user with printed logs of information on current or past states of the process.

Program Development:

1. Computer used: PDP-11/55, PDP1134
2. Minimum core: 16 K words.
3. Operating environment: Real time (RSX11M)
4. Language: FORTRAN IV
5. Typical Run time: 5 minutes (depends on process scale)
6. Number of source cards: 500

PLC: Plant Control Program (Appendix A)

1. Introduction

Section 2 of this documentation is a general description of the PLC program—the functions it performs and its components. Section 3 discusses the preparation of the control subroutine that is required in order to use PLC. Section 4 covers the commands with which a user communicates with PLC, and section 5 explains the format of the logs produced by PLC. Section 6 explains the steps necessary to initiate PLC from the user's terminal.

2. General Description:

A *task* on the PDP-11 is an independently executable program. Put differently, a set of tasks is a set of programs which can run "simultaneously" (of course programs can't really run simultaneously, since there is only one central processor in the computer, but a set of tasks can appear to run simultaneously when the computer shares time among them in a very rapid round robin fashion). The Plant Control Program (called PLC) consists of three tasks which often are running "simultaneously." They are the Master task (henceforth called M); the Control task (henceforth called C), and the logging task (henceforth called L). M, as its name suggests, is the top level director of the plant, and also the handler of the user's requests. All of the user's communication with the computer when PLC is operating is via commands given to M. M is always active. C is the task which actually monitors and adjusts the state of the plant, utilizing a user written subroutine (henceforth called the CNTL subroutine) containing a control algorithm. C is activated periodically at a frequency specified by the user. In addition to its control function, each time it is activated, C writes a record to a "history file," recording the state of the plant at that time. L is also activated periodically at a frequency specified by the user, and writes a record of the state of the plant at that time to the Decwriter.

Special conditions which arise in the plant are reported on the Decwriter; specifically, actual pressures being outside certain limits or CNTL functioning incorrectly. These conditions will be discussed more fully.

3. Preparation of the CNTL Subroutine:

The CNTL subroutine is a user-written subroutine for carrying out the main control function, that is, for the selection of pressure set points (in Supervisory Mode) or control valve positions (in Direct Digital Control).

Log on to your own UIC. All the information CNTL requires to compute a control decision, and the decision arrived at, are communicated through common variables, thus it has no input or output parameters. Its first line is therefore:

SUBROUTINE CNTL

(Note: It must be named CNTL). Comments may follow this first line, but the first noncomment lines which follow it must be created as follows—enter EDI Command Mode and issue the command:

*UNSAVE DK1: [100,1]CO,M2.FTN

This will add the necessary common variables to the program. Then issue an I command to return to Input Mode and enter the rest of your program. Table 1 lists the actual lines that should be added to your program by the UNSAVE command. See also the following sample CNTL subroutines.

Table 1.

```
COMMON /FLAGS/ IFLG(2),NUMFLG
COMMON /CHAN/ ICHAD(2),ICHDA(5),NCHAD,NCHDA
COMMON /DATEL/ AD(2),DA(5)
COMMON /LUN/ LUNAD,LUNTI,LUNPLT,LUNDMP
COMMON /CALIB/ A(3),PUNITS,FUNITS,NUMCAL
COMMON /BOUNDS/ PHALM(2),PLALM(2),PEMG(2),NUMALM
COMMON /GOALS/ PRODR,PRODO,PRATIO
COMMON /MODE/ CURMOD,DIRDIG,SUPERV
COMMON /TIMINT/ CNTDLY,LOGDLY,CNTUN
COMMON /ACTUAL/ P(2),F
COMMON /CNTRL/ DV(2)
COMMON /OBJCTV/ SET(2)
COMMON /CNTLC/ ICOUNT,CK(2),TI(2),TD(2)
COMMON /SPACE/ ITOP,IMAX,LREC(220,6),RREC(220,7)
REAL*S PUNITS,FUNITS
INTEGER PRODO
LOGICAL*1 CURMOD,DIRDIG,SUPERV
INTEGER CNTDLY
LOGICAL*1 CNTUN
LOGICAL*1 LREC
```

The common variables which a CNTL subroutine may want to make use of will now be described. Many of these are arrays, and will be written here as they would appear in an array declaration. For example, A(3) in Table 2 means that an array of length 3 is in common, and a user's program may therefore make use of the array elements A(1), A(2), and A(3).

The values of all the common variables in Table 2 are defined by the user via Master commands. See Section 4 for a description of these commands.

The values of the following common variables have been newly computed by PLC each time CNTL is called:

P(2)	Real	P_1 and P_2
F	Real	Flow rate

The following variable may optionally be made use of by the user. It is set to zero when PLC is started and incremented by one each time CNTL is called:

ICOUNT Integer Number of times CNTL has been called

If the user wishes to use the time of day in CNTL, it may be obtained by:

CALL TIME(IH, IM, IS, IT)

IH will contain the hour (0 – 23), IM the minute (0 – 59), IS the second (0 – 59), and IT the tick (0 – 59). A tick is 1/60 of a second.

CNTL must compute two new "control positions" and store them in the common variable:

DV(2) Real Control positions

Regardless of whether in Direct Digital or Supervisory mode, these control positions must be expressed as numbers in the range 0.0 to 1. An error message will be printed, and the control positions ignored, if the values of DV(1) or DV(2) lie outside this range.

Table 2.

VARIABLE	TYPE	DESCRIPTION
A(3)	Real	The quantities a_1, a_2, a_3 defined in Equation 6.
B(3)	Real	The quantities b_1, b_2, b_3 defined in Equation 6.
PUNITS	Alphanumeric	The units in which all pressures in PLC are expressed
FUNITS	Alphanumeric	The units in which all flows in PLC are expressed
PHALM(2)	Real	If P_1 exceeds PHALM(1) or P_2 exceeds PHALM(2) at any time, an alarm condition exists in the plant; the bell is rung and a message printed on the DECwriter
PRODR	Real	Current requested production flow rate
PRODO	Integer	Production objective: 1 = Minimum pressure 2 = Maximum pressure 3 = Constant pressure ratio
PRATIO	Real	If PRODO = 3, the value of the requested constant pressure ratio
CURMOD	Integer	Current mode: 1 = Direct Digital 2 = Supervisory
CK(2) TI(2) TD(2)	Real	Special parameters for CNTL. These can be assigned any values desired by the user with the S command.

Although many of the common variables added to CNTL by the UNSAVE command above are of no concern to a user, care must be taken not to employ a variable of the same name. Specifically, a user should not use any of the following names in CNTL for a local variable:

A,AD,B,CK,CNTDLY,CNTUN,CURMOD,DA, DIRDIG,DV,F,FUNITS,ICHAD,ICHDA,ICOUNT, IFLG,IMAX,ITOP,LOGDLY,LREC,LUNAD, LUNDMP,LUNTI,LUNPLT,NCHAD,NCHDA, NUMALM,NUMCAL,NUMFLG,P,PEMG,PHALM, PLALM,PRATIO,PRODO,PRODR,PUNITS,RREC, SET,SUPERV,TD,TI.

The sample subroutines are listed in Table 3.

4. Master Commands:

(The prompt character is $$)

Input commands:

The first group of commands described are those which ask for input from the user. In each such command, the name of the variable whose value is desired is displayed by PLC, followed by a question mark, followed by the Fortran format in which the input is to be given, e.g.

A(1)   F10.4

At the present time, erroneous format on input will usually abort PLC, so care should be taken. In the command explanations below, only variables not defined in Section 3 above will be described at length.

The first seven input commands (i.e., H,C,G,B,M,X,Y) are mandatorily executed without user intervention each time PLC is started, since the values of the variables inputted in these commands are needed in order for PLC to proceed. After this initial sequence has been completed, a $$ prompt

Table 3.

```
          SAMPLESUBPROGRAM FOR SUPERVISORY CONTROL
          SUBROUTINE CNTL
          DIMENSION PS(2)
          COMMON /FLAGS/ IFLG(2),NUMFLG
          COMMON /CHAN/ ICHAD(2),ICHDA(5),NCHAD,NCHDA
          COMMON /DATEL/ AD(2),DA(5)
          COMMON /LUN/ LUNAD,LUNTI,LUNPLT,LUNDMP
          COMMON /CALIB/ A(3),B(3),PUNITS,FUNITS,NUMCAL
          COMMON /BOUNDS/ PHALM(2),PLALM(2),PEMG(2),NUMALM
          COMMON /GOALS/ PRODR,PRODO,PRATIO
          COMMON /MODE/ CURMOD,DIRDIG,SUPERV
          COMMON /TIMINT/ CNTDLY,LOGDLY,CNTUN
          COMMON /ACTUAL/ P(2),F
          COMMON /CNTRL/ DV(2)
          COMMON /OBJCTV/ SET(2)
          COMMON /CNTLC/ ICOUNT,CK(2),TI(2),TD(2)
          COMMON /SPACE/ ITOP,IMAX,LREC(220,6)RREC(220,7)
          REAL*S PUNITS,FUNITS
          INTEGER PRODO
          LOGICAL*1 CURMOD,DIRDIG,SUPERV
          INTEGER CNTDLY
          LOGICAL*1 CNTUN
          LOGICAL*1 LREC
C         COMPUTE DESIRED PRESSURE P2 TO SATISFY PRODUCTION RATE
          W=PRODR/A(3)
          PS(2)=(W-10.)/SORT(0.47)
          IF(PS(2).LT.13.)PS(2)=(SORT(4.*W**2+14.7**2)-14.7)/2.
C         COMPUTE DESIRED PRESSURE P1 ACCORDING TO DESIRED DISTRIBUTION
C         1-FOR MIN P1,2-FOR MAX P1,3-FOR CONST P1 *PRATIO=P1/P2
          IF(PRODO,NE.3)GO TO 510
          PS(1)=PS(2)*PRATIO
          GO TO 540
510       PS(1)=60
          IF(PRODO.NE.1) GO TO 540
          IF((P2).LT.PS(2)).OR.((P(1)-PS(2)).LE.1.))GO TO 540
          DV(1)=DV(1)-TD(1)*(P(1)-P(2))
          GO TO 530
540       CONTINUE
C         COMPUTE TRANSMITTED SET POINTS
C         (COMPENSATION ALGORITHM)
          DV(1)=PS.1)/60.
530       DV(2)=PS(2)/60.
C     SATURATION POSITIONS
          IF (DV(1).GT.1.) DV(1)=.99
          IF (DV(2).GT.1.) DV(2)=.99
          IF (DV(1).LT.0.) DV(1)=.01
          IF (DV(2).LT.0.) DV(2)=.01
          RETURN
          END
```

appears, and the user can give any command (including another H,C,G,B,M,X, or Y).

i) H command (channels):
ICHDA(5) and ICHAD(2) are inputted.
ICHDA(5) are the actual D/A channels corresponding to DA1,DA2,DO1,DO2, and DO3 in Figure 1.
ICHAD(2) are the actual A/D channels corresponding to AD1 and AD2.

ii) C command (Calibration):
A(3),B(3),FUNITS, and PUNITS are input.

iii) G Command (Goals):
PRODR,PRODO, and PRATIO are input.

iv) B command (Bounds):
PHALM(2) is input.

v) M command (Mode):
Select Direct Digital or Supervisory control.
CURMOD is input.

vi) X command (Set delay between logs):

The variable LOGDLY, whose value is the number of seconds between logs is input. Thus the value of LOGDLY determines how often L is activated. The format of a log is described in Section 5.

vii) Y command (Set delay between control decisions):
The variables CNTUN, specifying a time unit of ticks or seconds (a tick is 1/60 second), and CNTDLY are input. The value of CNTDLY is the number of ticks or seconds between calls to CNT1 (and between records of plant state being written to the History File). Thus the value of CNTDLY determined how often C is activated.
Note: giving a negative value to LOGDLY or to CNTDLY will cause there to be no further activations of (respectively) L or C.

viii) S command (Special parameters for CNTL):
CK(2),TI(2), and TD(2) are input. The user can assign these any values desired and then make use of them in CNTL.

Miscellaneous commands

ix) L command (Log - special):
Print a log (in the same format as a regular log) on the Decwriter, immediately. This command has not effect on the regular logs requested with the Y command.

x) D command (Dump):
Dump the contents of the History File on the line printer. Not intended for general use.

xi) I command (Immediate D/A):
Place a specified voltage (between 0. and 10. volts) on a specified D/A channel immediately. A D/A channel specified here is any actual channel (i.e. in the range 1-16), independent channel assignments made with the H command.

xii) R command (Reset):
Erase the contents of the history File and set ICOUNT (cf. Section 3) to 0.

xiii) P command (Plot):
Plot the History File. all the History File values of any of 5 variables, namely P(2), SET(2) (see Section 5 below for the definition of SET(2)), and F may be plotted; up to 3 of these may be plotted on the same graph.

xiv) E command (Exit):
Shut off the plant and exit from PLC.

5. Logging

An example of a typical log is the following:

3:38:46 9 CONSTPR SUPERVIS
P(1)=4.2062 SET(1)=3.2062 P(2)=4.1866
SET(2)=5.000
FA=4.1866 FR=3.000
REQUESTED CONSTANT PRESSURE
RATIO=2.000

Here 3:38:46:9 is the time of day in hours:minute:second:ticks. CONSTPR (i.e. constant pressure) is the value of PRODO (MINPRES or MAXPRES could also appear here). SUPERVIS is the mode (DIR DOTL could also appear here).

SET(1) and SET(2) are the last control positions computed by CNTL, expressed in engineering units. They are in pressure units, if in supervisory mode, and as percentages, i.e. numbers in the range 0. to 100., if in direct digital mode. Since in the log above we are in supervisory mode, they are expressed as pressures.

FA (Flow actual) is the value of F; FR (Flow requested) is the value of PRODR. Since PRODO is constant pressure ratio, the last line gives the value of PRATIO.

6. Steps for the use of PLC

Be sure you're at a Tektronix terminal, not the Decwriter, as the Decwriter is used as a console device.

i) Prepare the CNTL subroutine as described in Section 3, on your own UIC.
ii) Compile CNTL. Correct compilation errors and recompile until the subroutine is error-free.
iii) Set the UIC to [100,1]
iv) Issue the command

PIP DK1:=DK1:[100,iii]CNTL.FTN/NV

where [100,iii] is your UIC (see Manual p.6-11 for meaning of /NV)

v) Type the following MCR command:

@DK1:BEGIN

This will take about five minutes to execute.
vi) Type the following MOR command:

@DK1:PLC

You will now be talking to M.
vii) If you exit from PLC, and then just want to run it again, repeat step vi.
viii) If your CNTL subroutine contains errors and you want to correct them, set the UIC to your own

code, correct the errors in EDI, then repeat steps ii-vi, except at step v type the following in place of what is given at step v above:

@DK1:CNTBLD

ix) If you want to make a completely new CNTL subroutine, repeat steps i-vi, except at step v type instead the following:

@DK1:CNTBLD

x) If the computer crashes, when it is again, type:

@DK1:BEGIN2

and when this has finished.

@DK1:PLC

to restart PLC.
xi) When you have finished using PLC and are ready to sign off the computer, type

@DK1:END

and when this has finished, the usual

@BYE

Module A4.3

Feedforward Control

T. F. Edgar

University of Texas
Austin, Texas

and

D. E. Seborg

University of California
Santa Barbara, California

OBJECTIVES

After completing this module, the student should be able to:

1. Understand the advantages and disadvantages of feedforward control (versus feedback control).
2. Know why feedback control must generally be used and how tuning of the feedback controller is affected by the presence of feedforward control.
3. Analyze a block diagram involving feedforward and feedback control.
4. Derive and apply the formula for static and/or dynamic feedforward compensation.
5. Use physical models (material or energy balances) in conjunction with feedforward control.
6. Analyze the effect of time delays to determine whether feedforward control is realizable.

PREREQUISITE MATHEMATICAL SKILLS

1. Laplace transforms.
2. Transfer functions.
3. Block diagram algebra.
4. Use of deviation variables.

PREREQUISITE ENGINEERING AND SCIENCE SKILLS

1. Mass and energy balances.

Feedforward control has generally been considered to be an advanced control method which can overcome many of the disadvantages of feedback control. Feedback (FB) is a proven process control technique which is widely used in the process industries. Some of the reasons feedback control has assumed a dominant position are:

a) Corrective action occurs regardless of the source and type of disturbance.

b) FB control requires little knowledge about the process, i.e., it does not require an explicit process model since tuning can be performed on-line.

c) A feedback controller can be made versatile and robust through proper design.

The general structure of a FB system is shown in Figure 1; L is the load variable, U the manipulated variable, and C is the controlled variable. With feedback control one desires to maintain C at its set point, C_{set}, despite load disturbances.

However, some of the major disadvantages of feedback control are listed below:

a) Feedback control takes no corrective action until a deviation in the controlled variable (C) is measured. One important consequence of this feature is that no predictive action on the part of the controller can be taken. FB control is not capable of achieving "perfect control."

b) Feedback control may perform poorly for processes with large process lags or time delays between the controller (U) and output (C). Figure 2 shows the effectiveness of a three mode controller for such a system.

c) Frequent and severe disturbances may cause the process to operate continually in a transient state, i.e., the process may not settle out.

These disadvantages make feedforward control an attractive alternative for some processes.

The philosophy of feedforward (FF) control is different from feedback control. The key feature of FF control is that L, the disturbance, is measured, and U is adjusted in order to maintain C at C_{set}. This

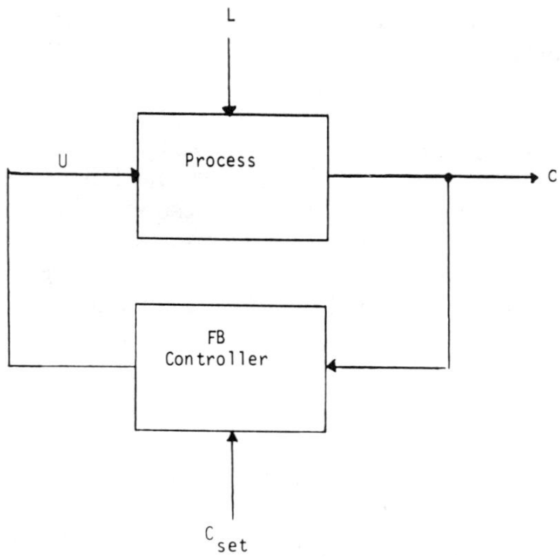

Figure 1. Feedback control structure.

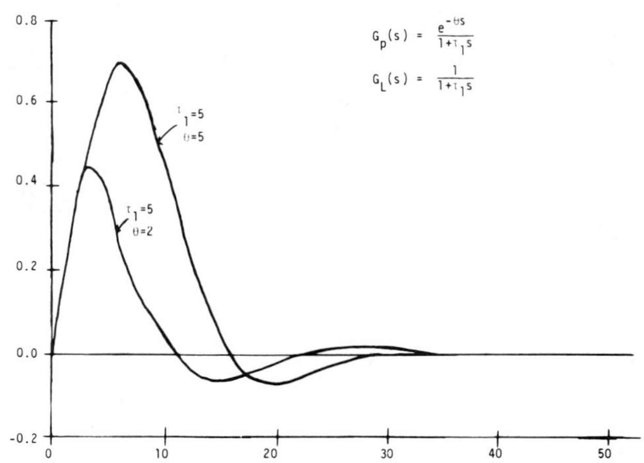

Figure 2. Comparison of the effect of a step disturbance on the response for two dead times.

approach is illustrated by Figure 3; note that the controlled variable is not measured. The advantages of FF control are as follows:

a) The FF controller takes corrective action when the disturbance is measured, or before the process is upset. Because the controller can act before the disturbance affects the process, perfect control is theoretically attainable.

b) The FF controller does not affect system stability and enhances the effectiveness of an existing FB system.

On the other hand, there are several undesirable features of FF control:

a) The disturbance must be measured on-line, which may not be possible or economical.

b) A reasonably accurate process model is required for effective use of FF control; an inaccurate process model may negate the potential improvements obtainable through FF.

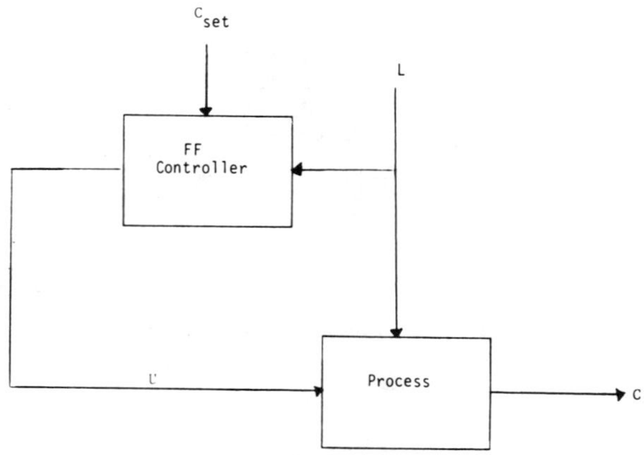

Figure 3. Feedforward control structure.

c) The ideal feedforward controller may be physically unrealizable. However, approximations to the ideal controller still may provide very good control.

One obvious solution to the FF/FB dichotomy is to use the two methods simultaneously. FF control therefore would attempt to eliminate the effects of measurable disturbances, while FB control would correct for unmeasurable disturbances and modeling errors. This is often referred to as feedback trim. These controllers have become accepted practices in the chemical process industries since the 1960's [Shinskey (*1*)].

FF DESIGN BASED ON MATERIAL AND ENERGY BALANCES

Consider a heat exchanger example (see Figure 4) to illustrate the use of FF and FB control. The control objective is to maintain T_2, the exit liquid temperature, at the desired value (or set point), T_{set}, despite variations in inlet liquid flow rate, w, and inlet liquid temperature, T_1. This is done by manipulating w_s, the steam flow rate. A feedback control scheme would entail measuring T_2, comparing T_2 to T_{set}, and then adjusting w_s. A feedforward control scheme requires measuring w and T_1, and adjusting w_s (knowing T_{set}), in order to control exit temperature, T_2.

Figures 5a and 5b show the control system diagrams for FB and FF control. A feedforward control algorithm can be designed for the heat exchanger in the following manner (*1*). Using a steady-state energy balance and assuming no heat from the heat exchanger.

$$w_s \Delta H_v = w C_p (T_2 - T_1) \tag{1}$$

where

ΔH_v = latent heat of vaporization

C_p = specific heat of liquid

Rearrange Equation 1,

$$w_s = \frac{C_p}{\Delta H_v} w(T_2 - T_1) \tag{2}$$

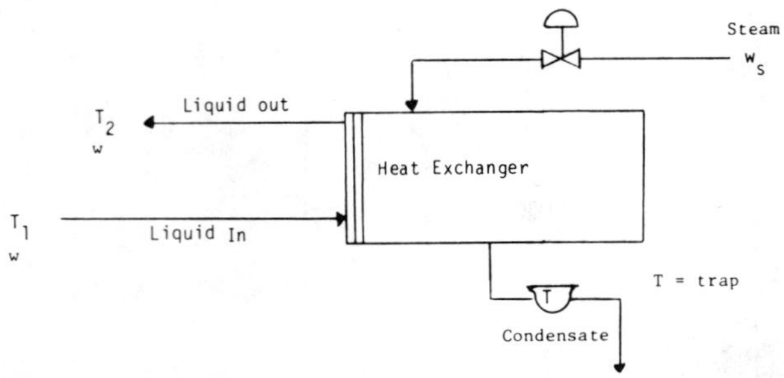

Figure 4. A heat exchanger diagram.

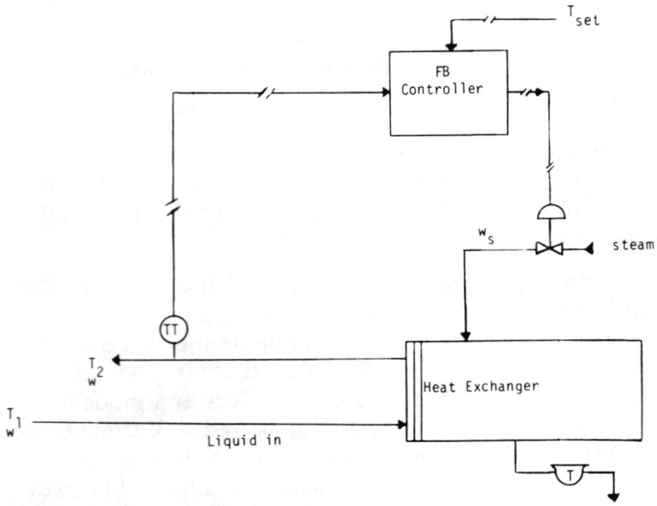

Figure 5a. Feedback control of a heat exchanger.

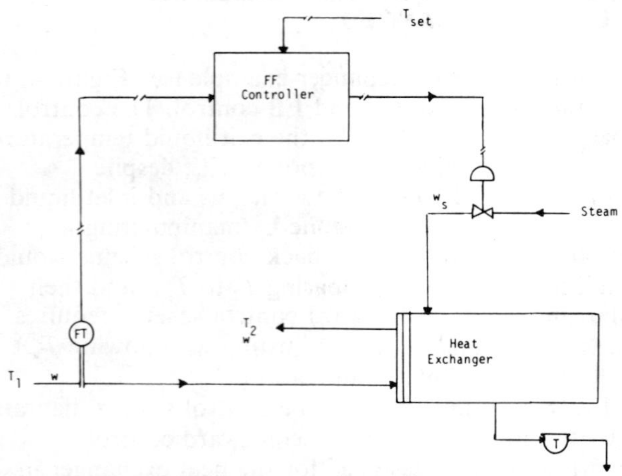

Figure 5b. Feedforward control of a heat exchanger.

or

$$w_s = Kw(T_2 - T_1) \tag{3}$$

with

$$K = \frac{C_p}{\Delta H_v} \tag{4}$$

Replace T_2 by T_{set}:

$$w_s = Kw(T_{set} - T_1) \tag{5}$$

Equation 5 can be used in the FF calculation, assuming one knows the physical properties C_p and ΔH_v. Of course, it is probable that the model will contain errors (e.g., unmeasured heat losses, incorrect C_p or ΔH_v). Therefore, K can be designated as an adjustable parameter which can be tuned. The use of a physical model for FF control is desirable since it provides a physical basis for the control law and gives an a priori estimate of what the tuning parameters are. Note that such a model could be nonlinear (e.g., in Equation 5, w and T_{set} are multiplied).

Block Diagram Analysis

One shortcoming of this feedforward design procedure is that it is based on the steady-state characteristics of the process and as such, it neglects process dynamics; i.e., how fast the controlled variable responds to changes in the load and manipulated variables. Thus, it is often necessary to include "dynamic compensation" in the feedforward controller. The most direct method of designing the FF dynamic compensator is to use a block diagram of a general process, as shown in Figure 6, where G_T represents the disturbance transmitter, G_F is the feedforward controller, G_L is the load, G_v the valve, and G_p, the process. G_m is the output transmitter and G_c is the feedback controller.

Using block diagram algebra, the output $C(s)$ is given by

$$C(s) = \frac{G_T G_F G_v G_p L(s) + G_L L(s)}{1 + G_m G_c G_p G_v} \tag{6}$$

$C(s) = 0$, or zero error, is desirable for the disturbance rejection $[L(s) \neq 0]$. Solving Equation 6 for G_F,

$$G_F = \frac{-G_L}{G_T G_v G_p} \tag{7}$$

Example

Let us now design a feedforward controller with dynamic compensation, assuming that the dynamics of G_L and G_p are first order; in addition, assume that $G_m = G_v = G_T = 1$ (for simplicity).

$$G_L(s) = \frac{K_L}{\tau_L s + 1} = \frac{C(s)}{L(s)} \tag{8}$$

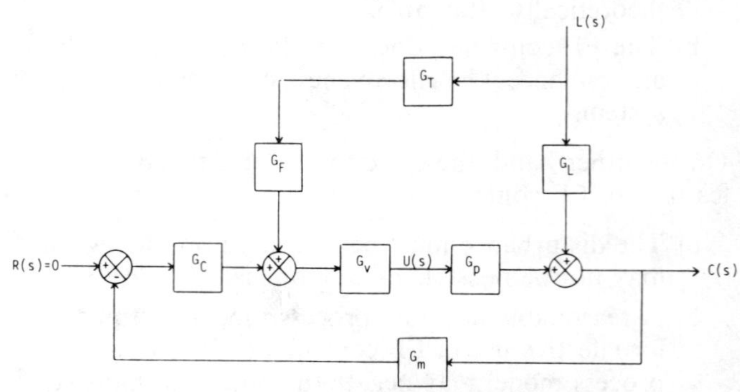

Figure 6. Block diagram for feedback-feedforward control.

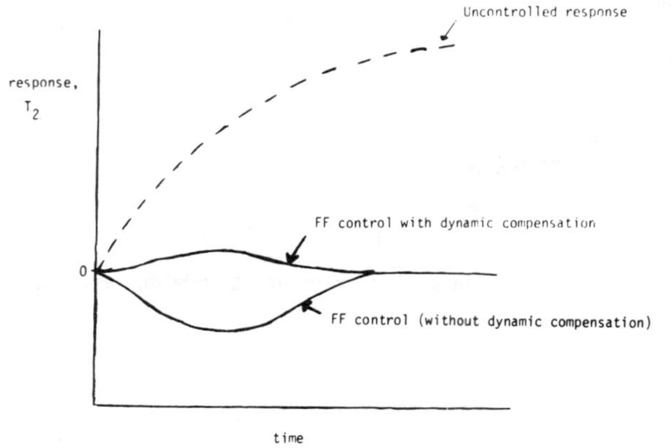

Figure 7. Effect of dynamic compensation on feedforward control for a step disturbance in the load variable.

$$G_p(s) = \frac{K_p}{\tau_p s + 1} = \frac{C(s)}{U(s)}$$

Solution:

Using Equation 7,

$$G_F(s) = -\frac{\tau_p s + 1}{\tau_L s + 1} \times \frac{K_L}{K_p} \tag{9}$$

In Equation 9, a physical model can be used to compute K_L/K_p. For example, the steady-state gains K_L and K_p for the heat exchanger system discussed earlier can be formed by rearranging Equation 3 to give:

$$\bar{T}_2 = \bar{T}_1 + \frac{\bar{w}_s}{K\bar{w}} \tag{10}$$

$$K_p = \left(\frac{\partial \bar{T}_2}{\partial \bar{w}_s}\right) = \frac{1}{K\bar{w}} \tag{11a}$$

where the bar over a variable denotes a steady-state value. If the load disturbance is the inlet temperature, then

$$K_L = \left(\frac{\partial \bar{T}_2}{\partial T_1}\right) = 1 \tag{11b}$$

Substitution into Equation 9 gives

$$G_F(s) = -\frac{\tau_p s + 1}{\tau_L s + 1} \times K\bar{w} \tag{12}$$

The FF controller is in this case a simple "lead/lag unit" which can be implemented using analog elements, or by a digital computer (*1*). Figure 7 shows typical responses for steady-state versus dynamic FF compensators for the case where simple dynamic models are an approximation to the true dynamics. In practice, the engineer can "tune" $K\tau_p$ and τ_L in the field to improve the performance of the FF controller.

The feedforward controller can also be simplified to provide steady-state feed forward control. This is done by setting $s=0$ in $G_F(s)$. This might be appropriate if there is uncertainty in the dynamic models for G_L and G_p.

Blending Control

One application where feedforward control is beneficial is in blending operations, the dynamics of which are often dominated by transport lag, or dead time. Clearly it would be undesirable to wait until after the stream is in the storage tank to measure the composition; rather, one would want to detect a change in a feed stream rate or composition before the stream enters a storage tank. Figure 8 illustrates a typical blending operation. If stream 1 is the disturbance stream and stream 2 is the control stream (which is adjusted to compensate for variations in stream 1), it is clear that stream 2 must be located after the measurement point in stream 1, or else the FF scheme will not be able to obtain the necessary information to counteract the disturbance. Assume the load (disturbance) variable is x_1, the manipulated variable is F_2, F_1 is a constant, and the controlled variable is x_3. Suppose that the load transfer function is $G_L = K_L e^{-\theta_L s}$ (a pure time delay model). In addition, assume the process transfer function is $G_p = K_p e^{-\theta_p s}$. The FF controller for $G_T = 1$ and $G_v = 1$ is therefore,

$$G_F = \frac{-G_L}{G_p} = \frac{-K_L}{K_p} e^{-(\theta_L - \theta_p)s} \tag{13}$$

Note that if the net dead time term, $(\theta_L - \theta_p)$, is positive, G_F is realizable. However, if $\theta_L < \theta_p$ (the transport delay between the analyzer and the storage tank is less than the delay between the control valve and the storage tank), the FF computer has a positive argument in the exponential, signifying that the computer must predict the disturbance before it is measured. This obviously is an unrealizable controller, thus successful FF control always requires that the controller effectively "intercept" the disturbance before the output variable exhibits large errors.

Distillation Control

Feedforward control of distillation columns has been an important concern in process plants (*1*). The importance of close control of distillation columns is summarized below:

1. Product is at terminal stage of refinement and is most valuable at this point.
2. Product quality specifications *must* be met.

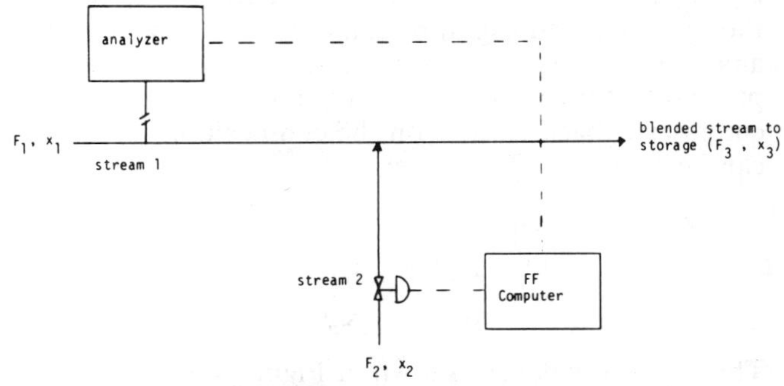

Figure 8. Blending operation diagram.

Difficulties in control of columns include such factors as:

1. Towers with many trays have slow responses, often with large time delays.
2. Separation is affected by many variables.
3. Control loops interact.
4. On-line analysis for product composition is not always feasible.
5. Distillation units are usually last in the processing sequence, and thus are subject to disturbances from upstream units.

Consider the case of a disturbance in the feed flow rate, F. It would be desirable to design a FF controller to handle these disturbances, since feedback control yields long recovery times. Here, the distillate flow rate is selected as the manipulated variable (1). If F is measured and X_D and X_B (the distillate and bottoms composition, respectively) are set at desired values, steady-state distillate flow (D) can be computed by a material balance (as the following discussion explains). However, the effect of dynamics between D, F and X_D must also be considered. In a study at Exxon (3),

$$G_L(s)=\frac{\Delta X_D(s)}{\Delta F(s)}$$

$$=\frac{e^{-30s}}{1+60s}\times K_L=\frac{C(s)}{L(s)} \quad (14)$$

$$G_p(s)=\frac{\Delta X_D(s)}{\Delta D(s)}$$

$$=\frac{e^{-20s}}{1+95s}\times K_p=\frac{C(s)}{U(s)} \quad (15)$$

where Δ denotes a deviation variable and the process gains are unspecified. Note that the dead time of the disturbance is greater than the dead time for control, so FF control is feasible.

$$\frac{G_L(s)}{G_p(s)}=\frac{e^{-10s}(1+95s)}{1+60s}\times\frac{K_L}{K_p}$$

$$=\frac{U(s)}{L(s)} \quad (16)$$

Therefore, D is computed from F using the above formula. Pure dead time cannot be simulated using analog components (it can be on a digital computer, though), and lead–lag approximations must be used for analog control (1). The pure dead time expression is preferable if it can be implemented. One can also employ feedback control on the composition X_D, if it can be measured. Therefore:

$$\Delta D(s)=G_F\,\Delta F(s)$$

$$=-\frac{e^{-10s}(1+95s)}{1+60s}\times\frac{K_L}{K_p} \quad (17)$$

The material balances shown in Figure 9 can be used to find the ratio of K_L/K_p; the steady state equations

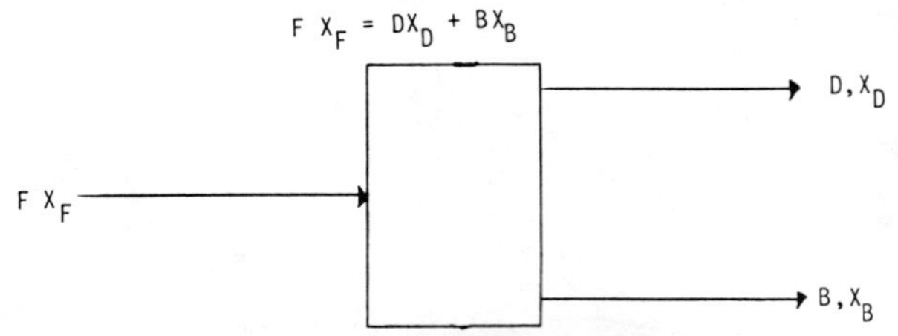

Figure 9. Material balance for a distillation column.

are:

total material balance: $F=D+B$ (18)

component balance: $FX_F=DX_D+BX_B$ (19)

where X_F is the feed composition and B is the bottoms flows. Solving for D,

$$D=\frac{F(X_F-X_D)}{(X_D-X_B)} \quad (20)$$

Next, calculate the ratio of K_L and K_p.

$$K_L/K_p=\left(\frac{\partial X_D}{\partial F}\bigg/\frac{\partial X_D}{\partial D}\right)_{ss}$$

$$=\left(\frac{\partial D}{\partial F}\right)_{ss}=\frac{X_F-X_D}{X_D-X_B} \quad (21)$$

Note that the gain ratio can be updated as a function of the product and feed specifications.

Other Considerations in Feedforward Control

The tuning of feedforward and feedback control systems does not depart from methods used to tune FF and FB systems independently. In analyzing the block diagram in Figure 6, note that G_F is chosen to cancel out the effects of the disturbance, $L(s)$; hence, in theory $C(s)$ is not affected by $L(s)$, as long as there are no model errors. For the feedback loop, therefore, the effects of $L(s)$ can also be ignored, which for the servo case is:

$$\frac{C(s)}{R(s)}=\frac{G_cG_vG_p}{1+G_cG_vG_pG_m} \quad (22)$$

Note that the characteristic equation will be unchanged for the FF+FB system, hence system stability will be unaffected by the presence of the FF controller. Of course, one must expect that with feedforward control, the tuning of the FB controller can be less conservative than that for the case of FB alone, since smaller excursions from the set point will result, which would in turn, make the dynamic model, $G_p(s)$, more accurate.

NOMENCLATURE

B = bottoms molar flow rate, distillation column
c, C = controlled variable
C_p = specific heat
C_{set} = set point for controlled variable
D = distillate molar flow rate

F	= feedflow rate
G_c	= feedback controller transfer function
G_F	= feedforward controller transfer function
G_L	= load transfer function
G_m	= measurement transfer function
G_p	= process transfer function
G_T	= load transmitter transfer function
G_v	= valve transfer function
K	= controller tuning parameter
K_L	= gain of load transfer function
K_p	= gain of process transfer function
L	= load input or disturbance
s	= Laplace transform variable
T_1	= inlet temperature in heat exchanger
T_2	= Outlet temperature in heat exchanger
T_{set}	= set point for T_2
w	= cold-fluid flow rate in heat exchanger
w_s	= steam flow rate in heat exchanger
x_1, x_2, x_3	= compositions
X_B	= bottoms mole fraction
X_D	= distillate mole fraction
X_F	= feed mole fraction
τ	= time constant
θ	= time delay
ΔH_v	= heat of vaporization
ΔX_B, ΔX_o	= deviation variables

LITERATURE CITED

1. Shinskey, F. G., "Process Control Systems," 2nd Ed., McGraw-Hill, New York (1979).
2. Smith, C. L., "Digital Computer Process Control," Intext, PA (1972).
3. MacMullan, E. C., and F. G. Shinskey, "Feedforward Analog Computer Control of a Superfractionator," *Control Eng.*, (March 1964).

STUDY PROBLEMS

1. Contrast the key advantages and disadvantages of feedforward control and feedback control.

2. Describe the role of mathematical modeling in feedforward controller design. Discuss how a black box model can be used for static feedforward control and for dynamic feedforward compensation.

3. Are the contributions of feedforward and feedback control always additive?

4. How does the inclusion of feedforward control influence stability of the feedback loop?

5. Contrast the computational effort involved in design of FF control versus FB control.

6. Under what circumstances can FF control not be implemented physically?

7. Why are simple lead–lag networks sufficient for analog feedforward control when more complicated compensators are indicated by transfer function analysis? Does digital implementation remove restrictions imposed by analog devices?

HOMEWORK PROBLEMS

1. The transfer functions of a control system are

$$G_m = 1$$

$$G_v = 1$$

$$G_T = 1$$

$$G_L = \frac{2}{(s+1)(5s+1)}$$

$$G_p(s) = \frac{1}{s+1}$$

a) Design a feedforward controller with dynamic compensation and then design one without dynamic compensation (static FF).

b) Specify a feedback controller.

c) Simulate the performance of the dynamic and static FF controllers for a step load change, with no feedback control.

d) Repeat with proportional feedback control ($K_c = 5$).

2. Design a feedforward controller for

$$G_L = \frac{2}{(s+1)(5s+1)}$$

$$G_p = \frac{s+2}{(s+1)(5s+1)} \quad (G_m = G_v = G_T = 1)$$

Can you implement this controller (it is physically realizable)?

3. A distillation column is found to have transfer functions,

$$G_L(s) = \frac{X_D(s)}{F(s)} = \frac{K_L e^{-30s}}{1+60s}$$

and

$$G_p(s) = \frac{X_D(s)}{D(s)} = \frac{K_p e^{-20s}}{1+95s}$$

$$G_m = G_v = G_T = 1$$

Design an FF controller with dynamic compensation using a material balance model.

Module A4.4

Cascade and Ratio Control

T. J. Ward

Clarkson College of Technology
Potsdam, New York

OBJECTIVES

After completing this unit, the student should know what cascade and ratio control are and be able to recognize these schemes in typical control systems. In addition, he or she will have a list of references for additional study as needed.

PREREQUISITE MATHEMATICAL SKILLS

None.

PREREQUISITE ENGINEERING AND SCIENCE SKILLS

1. Single-loop feedback control concepts.

Cascade and ratio-control are two schemes that are widely utilized in the process industries. This unit will encourage familiarity with each in turn by looking at the configurations and some typical applications. Then, the unit will look at design procedures, tuning techniques, and hardware features.

CASCADE CONTROL

A. CONCEPT

In cascade control, two controllers are "cascaded" together so that the output of one is the setpoint input of the other. This is illustrated in the linear block diagram of Figure 1. There are other control configurations in which the controllers are cascaded, but the term is usually applied to control cases that have the two-loop control structure of Figure 1 and where both measurements are process variables or manipulative inputs (instead of disturbance inputs).

Based on this diagram, one can make the following observations:

1. Two process variables c_1 and c_2 are measured and used to actuate control. These are usually called the primary and secondary variables, respectively. The objective is the control of c_1.

2. There are two feedback loops. The outer loop is called the primary loop and the controller G_{C1} is

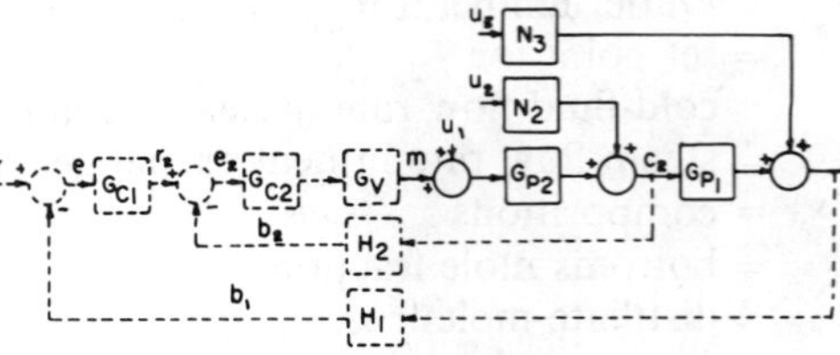

Figure 1. Cascade control block diagram.

called the primary controller. The inner loop is called the secondary loop and contains the secondary controller G_{C2}.

3. Only one process input is manipulated. As a result, there can only be one controlled variable. In other words, there is only one degree of operational freedom, as represented by the single setpoint input r.

4. The disturbances u_1, u_2, u_3, and u_4 are shown entering the linearized process model at four different locations so that one can consider the effect of disturbance location. For example, it should be obvious that the inner loop will respond to the disturbances u_1 and u_2 since they enter ahead of the secondary loop variable c_2.

The essential difference between this diagram and the more common single-loop feedback diagram is the introduction of the inner secondary control loop. Now just what is the purpose of the inner loop?

B. LOGIC

The secondary loop can be viewed in terms of four possible logic roles as follows (*16*):

1. *Inner Loop Compensation.* The secondary controller G_{C2} modifies the troublesome dynamics associated with the secondary process function G_{p2} and consequently simplifies the design of the outer primary controller G_{C1} (*8*).
2. *Dynamic System Modifier.* The two-loop combination can be designed to give overall dynamics that would be difficult to achieve with a single feedback controller (*11*).
3. *Partial Disturbance Compensation.* The secondary controller can initiate some manipulative effort to compensate for the u_1 or u_2 disturbance transmission before these disturbance effects reach c_1 (*12*).

4. *Primary Feedback.* The inner loop can be viewed as the primary feedback control, while the outer loop serves as a trimming or setpoint corrector for the inner loop. This approach, with either continuous or sampled outer loop measurement, is characteristic of some process cases. (The outer loop may be continuous, manual, or digital.)

While these seem to be inconsistent viewpoints, the control structure is equivalent in all cases. The suitability of a particular view depends on the process situation and requires some skill on the part of the designer. In any case, it is important to realize that the inner loop is an inherent part of the overall feedback control, as emphasized in role (2). As a result, the inner loop modifies the characteristic equation and changes the dynamic response of the process.

C. TYPICAL APPLICATIONS

Many cascade control applications have appeared in the literature. We will take a look at several of these in order to appreciate the wide applicability and diverse roles of cascade control.

Heat Exchanger Control

Consider the cascade control system shown in Figure 2. The primary controlled variable c_1 is the tube-side temperature, the secondary variable c_2 is the shell-side flow rate, and the manipulative input m is the control valve stem position. If the shell-side flow is a vapor flow, the control can be viewed as in the partial disturbance compensation role, where the inner flow loop regulates the flow in the face of line pressure changes quickly before these disturbances propagate through the temperature loop. This could also be viewed as a case of a valve having a pressure-dependent transfer function, as in inner loop compensation. Alternatively, the flow loop might be viewed as a primary control loop, with the outer loop providing a temperature trim to accommodate changes in the tube-side inlet temperature (role 4). In any case, the two-loop theme of the dynamic system modifier applies and is evident when transfer functions are written for the control system.

Accumulator Level Control

As a second example, let us look at Figure 3. The primary variable is the accumulator level c_1, the secondary variable is the overhead product rate c_2, and the manipulative input is the valve stem position m. Can this be interpreted in terms of the logic roles mentioned?

Bottoms Level Control

Another distillation example is shown in Figure 4. The reboiler (or sump) level is the primary variable, the bottoms flow is the secondary variable, and the control valve stem position is the manipulative input.

Furnace Temperature Control

Consider the gas-fired furnace shown in Figure 5. In this example, the primary variable is the product temperature c_1. The secondary variable is the gas flow rate, while the manipulative variable is the control valve stem position. Note how this responds rapidly to variations in gas pressure before such changes can affect the product stream.

Stirred-Tank Temperature Control

In all of the previous examples, the secondary variable is a flow. While this is quite common, it is also possible to have other types of secondary variables. The stirred tank example of Figure 6 has the jacket fluid temperature as the secondary variable. Note how the secondary loop can provide fast

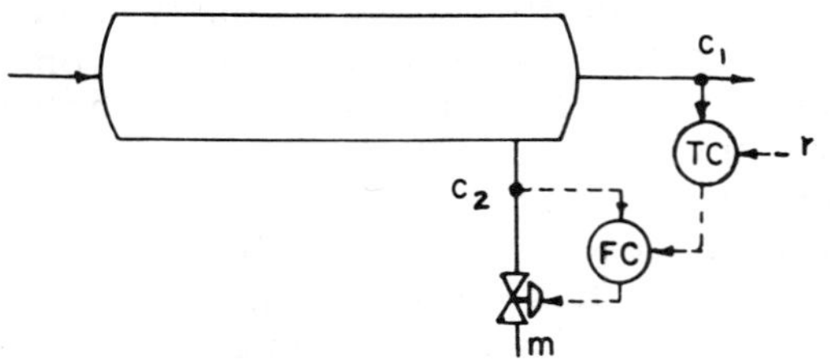

Figure 2. Heat exchanger control

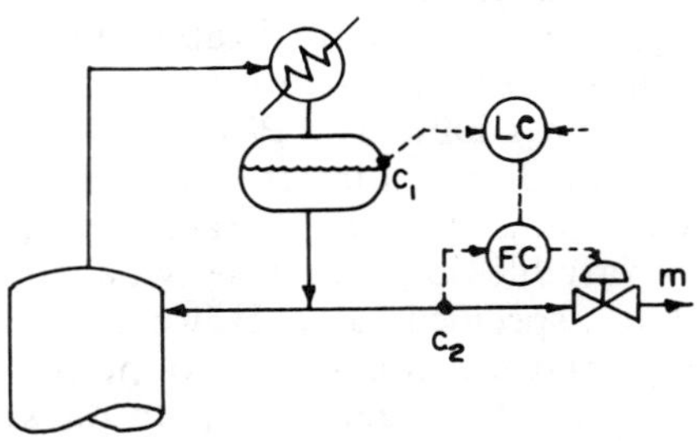

Figure 3. Accumulator level control.

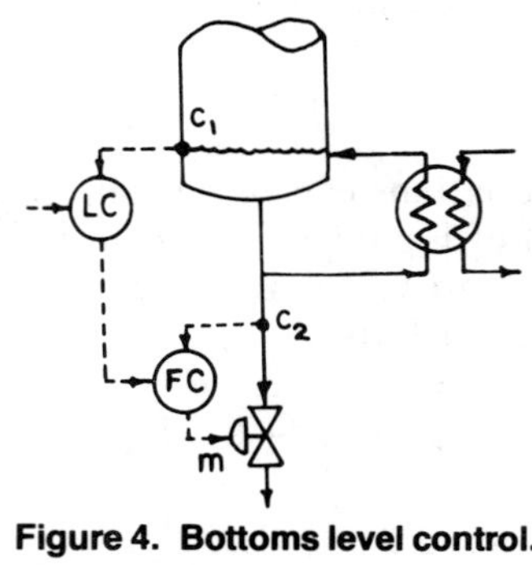

Figure 4. Bottoms level control.

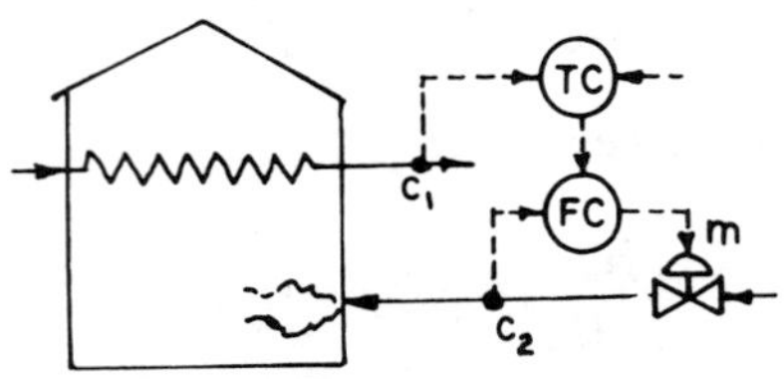

Figure 5. Furnace temperature control.

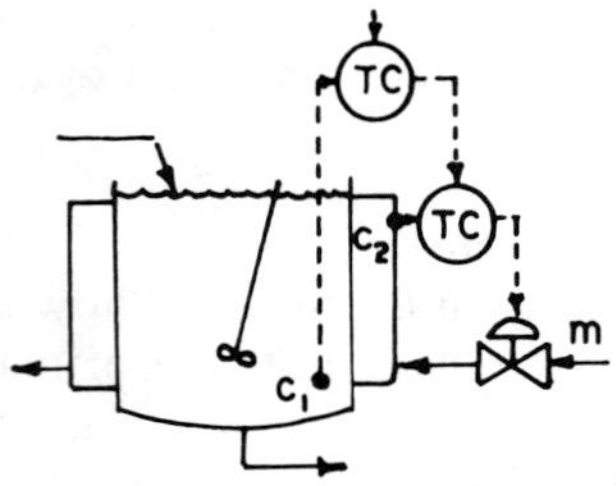

Figure 6. Stirred tank temperature control.

response to disturbances in the cooling water inlet temperature.

D. CASCADE CONTROL DESIGN

General

There seems to be some consensus in the literature that cascade control is more effective if a) the major disturbance enters the inner loop, and b) the inner loop is faster acting than the outer loop by a factor of two or three. However, there is also some improvement in response to disturbances that enter the outer loop because the cascade arrangement reduces the response time of the process elements in the inner loop (*10*, *13*).

The functional forms used for the two controllers could, in principle, be any realizable functions. Actually, most cascade systems use a proportional mode in the inner loop. The P-mode is used because a) the single mode has a faster response than the PI mode for the same stability characteristics, and b) the secondary variable does not usually have to be regulated at a specified valve. However, if the secondary variable contains a lot of noise, it might be desirable to add the integral mode to the secondary controller. The outer loop controller is usually a PI or PID controller. While the use of additional modes in both loops might be beneficial (*3*), the tuning difficulties tend to limit the controller types used in practice.

Analytical

Various equations can be written for the idealized linear model of Figure 1. For chemical process systems, the most useful are:

a) Inner Open Loop

$$b_2 = G_{p2} G_v G_{C2} H_2 e_2 + G_{p2} H_2 u_1 + N_2 H_2 u_2$$

b) Inner Closed Loop (Let $G_2 = G_{p2} G_v G_{C2}$)

$$b_2 = \frac{H_2}{1 + G_2 H_2} [G_2 r_2 + G_{p2} u_1 + N_2 u_2]$$

c) Overall Open Loop

$$b_1 = \frac{G_{p1} H_1}{1 + G_2 H_2} [G_2 G_{C1} e + G_{p2} u_1 + N_2 u_2] + N_3 H_1 u_3$$

d) Overall Closed Loop (Let $G_1 = G_{p1} G_2 G_{C1}$)

$$b_1 = \left(\frac{H_1}{1 + G_2 H_2 + G_1 H_1} \right) [G_1 r + G_{p1} G_{p2} u_1 + G_{p1} N_2 u_2 + N_3 (1 + G_2 H_2) u_3]$$

In many applications $G_{p2} = 1$, and $u_1 = u_3 = 0$. Further, it is occasionally possible to assume that $H_1 = H_2 = 1$. The stability of the model is determined by the roots of the characteristic equation of the overall closed-loop function.

Any conventional feedback design method can be applied if the model functions are well-defined. One example calculation (*4*) involves a) application of Bode open-loop criteria for inner loop design, b) closed inner-loop frequency response from a Nichols chart, c) evaluation of overall open-loop frequency response, and d) application of Bode open-loop criteria to overall open loop in order to design outer loop. While such methodology is useful for analysis, it is often not suitable for actual design. The reason, as noted by Douglas and others, is that the overall open-loop gain and phase curves are very steep in the neighborhood of the crossover frequency. If the transfer functions are not well-known or are likely to change over the operating range, the analytical designs would be of limited value.

Practical

As with many feedback systems, the design of cascade control typically involves little more than suitable selection of vendor hardware so as to match ranges and dynamic characteristics. These systems are field-tuned using classical tuning methods. The fact that extensive analytical efforts are not needed is a major reason why cascade control has found wide application in the process industry.

E. CASCADE CONTROL TUNING

The outer loop controller is first placed in the manual mode of operation. The inner loop controller is then tuned using any single-loop feedback tuning method. With the tuned inner loop in the automatic mode, the outer loop controller is then tuned. In some cases (*7*), it is then necessary to go back and reduce the inner loop gain in order to achieve good response.

F. HARDWARE

Conventional analog controllers can be used for cascade control. The only requirement is that the secondary controller have the capability of a remote setpoint input. This is generally available from most vendors. In addition, options are offered so that manual or automatic control of each loop can be set from a single control station.

Digital cascade control obviously is a simple implementation task. Nearly all digital computer control "packages" available from control system vendors include the cascade mode of operation.

RATIO CONTROL

A. CONCEPT

The purpose of ratio control is to control a variable relative to some other variable. In other words, the control maintains a desired "ratio" of the variables, rather than controlling the actual variables themselves. This is a desirable control configuration in blending operations, feeding reagents to a reactor in fixed proportions, and particularly in maintaining the fuel-air ratio in combustion processes. Ratio control can be represented by the block diagram of Figure 7. The external control configuration is quite similar to that of cascade control, but the purpose and logic are different. The following observations can be made from Fig. 7.

1. There are two process variables c_1 and c_2. These are often called the "controlled" and "wild" variables, depending on which one is affected by the manipulative input m.
2. The process variables c_1 and c_2 should have no dependence on each other. Further, only one should be affected by the manipulation m.
3. The ratio station is a logic function that multiplies a process variable by an adjustable ratio K.
4. The same objective can be accomplished more directly by using a divider function to operate directly on the two variables and feeding the output to the signal input of the controller. However, the divider would be inside the feedback loop. This would lead to a variable loop gain and might be troublesome (*13*).

As the student will see, the simple ratio control concept has wide application and is easily implemented.

B. CONTROL LOGIC

While the ratio concept is quite obvious, it is important to realize that neither variable is really controlled at a specified value. The block diagram of Figure 7 suggests that there is feedback control of the variable c_1. However, the variable c_2 is constantly changing the setpoint of the loop and c_1 is changing to maintain the c_1/c_2 ratio constant. It should be noted that the dynamic response of c_1 (and of the ratio) is affected by the form and parameters of the G_c controller.

If regulation of the two variables at specified values is desired, in addition to regulation of the ratio, then an additional feedback loop is needed for regulation of c_2. If the two variables are not both regulated, then it may be desirable to place high or low limiters on both variables. This may be necessary because regulation of only the ratio could result in excessively high or low values of one of the variables.

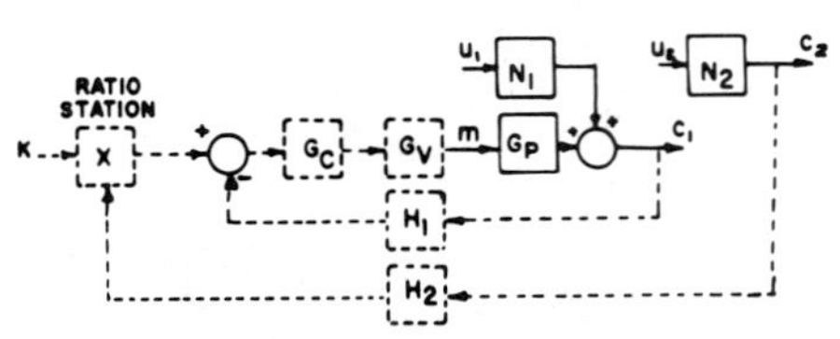

Figure 7. Ratio control block diagram.

C. TYPICAL APPLICATIONS

Ratio control is used primarily to maintain a constant flow ratio between two or more streams. Let us now consider various ways this has been utilized, as well as other applications.

Division Logic

As noted before, a division function can be used for ratio control. This is shown in Figure 8. Note that the division block is the apparent feedback loop for the flow c_1.

Ratio Station

The ratio station arrangement of Figure 7 can be used for ratio control. This is illustrated in Figure 9. The ratio station corresponds to an amplifier with an adjustable gain.

If orifice meters are used as flow detectors, then the flow transmitter output ΔP corresponds to the square of the flow. The ratio station setting must be the square of the desired ratio K. It is also possible to use a square root extractor unit between the flow transmitter and the ratio station.

Total Flow Control

The total flow of several regulated streams can be controlled and ratios maintained between individual streams. One open-loop arrangement for this purpose is shown in Figure 10. This uses a master set station for total flow. Since this is an open-loop system, it is susceptible to disturbance effects, such as pressure fluctuations in supply lines.

It may be preferable to close the loop by measuring total flow and adding a flow controller, as shown in

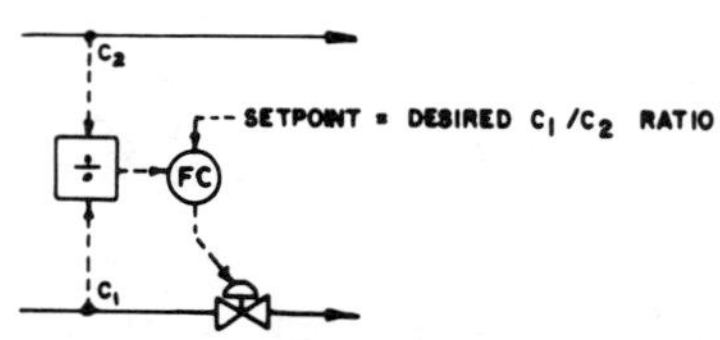

Figure 8. Ratio control using division element.

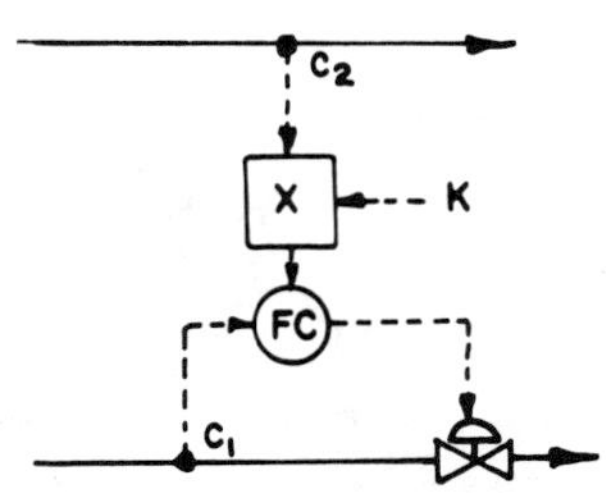

Figure 9. Ratio control using gain element.

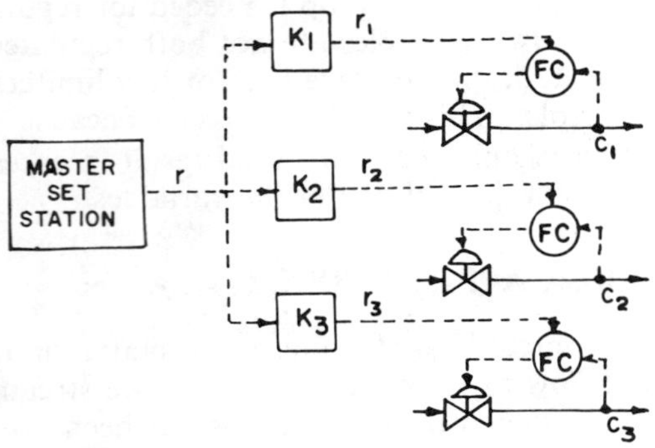

Figure 10. Total flow control—open loop.

Figure 11. Shinskey (*13*) cites use of this arrangement to control temperature in a reactor case where temperature is a direct function of total flow. Note that the overall configuration involves a cascade arrangement of the controllers. A third possibility would be to use a second flow feedback controller for the "wild" flow. This still may be subject to possible upstream disturbance effects.

Automatic Ratio Control

It is possible to use a closed-loop composition control to set the flow ratio automatically. This is useful in process operations where upstream disturbance effects might be troublesome. A typical arrangement is shown in Figure 12. The output of the c_1 composition controller is the ratio K. This is multiplied by the flow signal c_2 and used to set the flow controller setpoint r_3. Note that this also gives a cascade-like configuration.

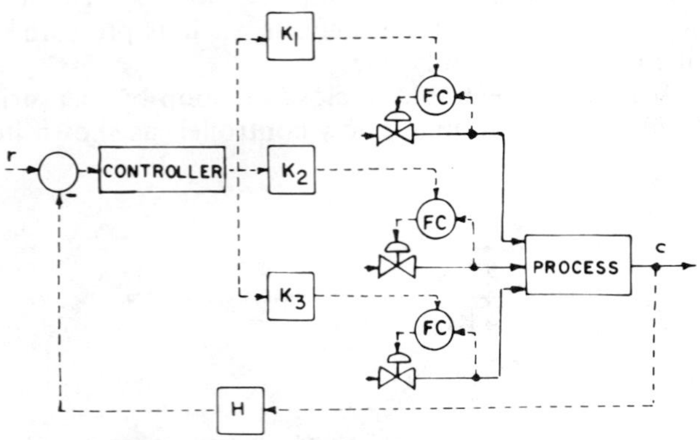

Figure 11. Total flow control—closed loop.

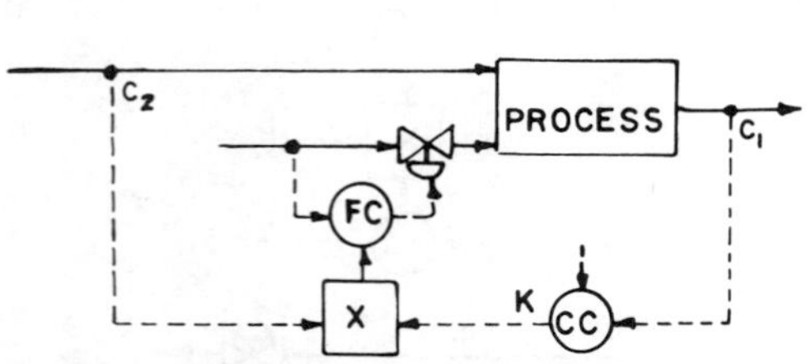

Figure 12. Automatic ratio control.

Digital Blending System

Various flow metering devices, such as turbine meters, are available with pulsed outputs. In such devices, the pulse rate is proportional to the volumetric flow rate. These can be compared with the pulsed setpoint signals in a up-down counter and the output used to drive a control valve. Many blending systems have such digital counters in ratio control configurations.

It should also be evident that any digital computer control scheme can readily be modified to include ratio control logic, as well as high and low limiters on either of the variables in the ratio.

Flue Gas Oxygen Control

In many combustion processes, the flue gas oxygen content is used as the controlled variable. This measurement is used to manipulate the fuel/air ratio in the burner feed. However, some limiter logic must be used to ensure that excess fuel does not accumulate in the furnace under all possible conditions of air flow. Shinskey (*13*) presents an interesting ratio control scheme for this case, as illustrated in Figure 13. Both flow controllers are set by the same master firing setpoint station. As a result of the arrangement, this scheme requires *identical* setpoint signals to each flow controller. The fuel/air ratio is changed to regulate the flue gas composition by using the product of air flow and oxygen composition as the input to the air flow controller.

If air flow is too low, its measurement is selected by the low selector as the setpoint for the fuel flow controller. If fuel flow is too high, its measurement is selected by the high selector as the setpoint for the air flow controller. This protects against high fuel and low air limits, as well as dynamic lags in setpoint response.

D. RATIO CONTROL DESIGN

There are few general guidelines available for ratio control design. However, the concept is simple and most difficulties are encountered in matching the ranges and input/output characteristics of the hardware items used. Any feedback controller can be field-tuned using conventional loop techniques.

CONCLUSION

Cascade and ratio control are schemes that have wide application in process control. Both are simple techniques for utilizing a second process measurement

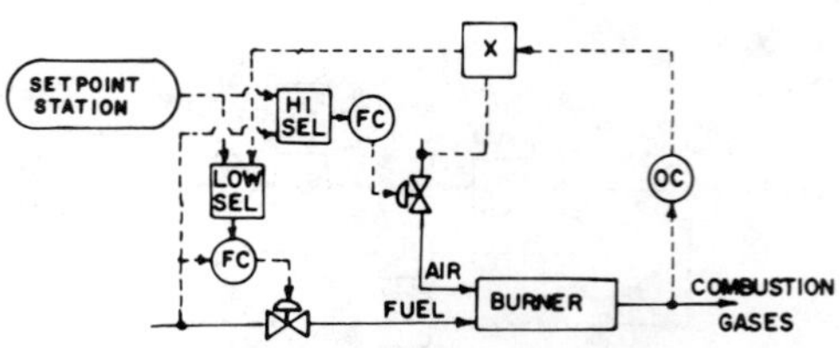

Figure 13. Flue gas oxygen control.

to achieve regulation objectives. Both can be implemented without extensive mathematical modeling or analysis. Further, both use readily-available hardware and can be field-tuned using empirical techniques. However, these methods require some skill and understanding by the designer.

LITERATURE CITED

1. Buckley, P., "Technique of Process Control," pp. 238–240, John Wiley, N.Y. (1964).
2. Chiu, K. C., Corripio, A. B., and C. L. Smith, "Tuning Cascade Control Systems," *Instr. & Control Systems, 47*, 1, 65 (Jan. 1974).
3. Clay, R. M., and C. D. Fournier, "Cascade vs. Single-Loop Control of Processes with Dead Time," *Proc. 1973 Joint Auto. Control Conf., 363*, IEEE, N.Y. (1973).
4. Douglas, J. M., "Process Dynamics & Control," Vol. 2, 175–184, Prentice-Hall, Englewood Cliffs, N.J. (1972).
5. Franks, R. G., and C. W. Worley, *Ind. Eng. Chem., 48*, 1074 (1956).
6. Gollins, N. W., *Control Eng., 3*, 7 (1956).
7. Harriott, P., "Process Control," pp. 154–166, McGraw-Hill, N.Y. (1964).
8. Koppel, L. B., "Intro to Control Theory with Applications to Process Control," pp. 25–27, Prentice-Hall, Englewood Cliffs, N.J. (1968).
9. Luyben, W. L., *Ind. Eng. Chem. Fund., 12*, 4, 463 (1973).
10. Murrill, P. W., "Automatic Control of Processes," pp. 431–440, Intext, Scranton (1967).
11. Perlmutter, D. D., "Intro. to Chemical Process Control," pp. 155–156, John Wiley, N.Y. (1965).
12. Pollard, A., "Process Control," pp. 351–352, American Elsevier, N.Y. (1971).
13. Shinskey, F. G., "Process Control Systems," pp. 154–167, McGraw-Hill, N.Y. (1971).
14. Webb, P. V., "Reducing Process Disturbances with Cascade Control," *Control Eng., 8*, 8, 73 (1961).
15. Wills, D. M., "Cascade Control Applications and Hardware", Bull, TX-119-1, Honeywell, Philadelphia (1960).
16. Yang, C. H., and Ward, T. J., *Ind. Eng. Chem. Fund., 14*, 3, 171 (1975).
17. Ziegler, J. G., "Cascade Control Systems," *Symp. Process Auto. Proc. Ind.*, Texas A&M, College Station (1954).

STUDY PROBLEMS

1. Derive the overall closed-loop equation for b_1 in the cascade control block diagram of Figure 1.

2. Explain why cascade control provides better response for disturbances that enter the inner loop of a cascade system.

3. Interpret the stirred-tank temperature control problem of Figure 6 in terms of the four logic roles given for cascade control.

4. What is the controlled variable in ratio control?

HOMEWORK PROBLEMS

1. Show how cascade control provides some improvement in response to disturbances that enter the outer loop by reducing the response time of the process elements in the inner loop. Hint: Assume an element $1/(Ts+1)$ and an inner loop proportional controller K.

2. Consider a special case of Figure 1 where $N_2 = H_1 = H_2 = G_v = 1$, $u_3 = u_1 = 0$. If both controllers are proportional ($G_{C1} = K_1$ and $G_{C2} = K_2$) and if the following are assumed,

$$G_{P2} = \frac{1}{(s+1)^3} \quad \text{and} \quad G_{P1} = \frac{1}{s+1}$$

then develop an analytical design based on the frequency response approach outlined in the module.

Module A4.5

Adaptive Control

T. F. Edgar

Department of Chemical Engineering
The University of Texas
Austin, TX 78712

and

D. E. Seborg

Department of Chemical Engineering
University of California
Santa Barbara, CA 93106

OBJECTIVES

At the conclusion of this module the student should be able to:

1. State the advantages of adaptive control over conventional control.
2. List some examples of programmed adaptation and when it can be used.
3. Understand the concept of self-tuning control and the role of parameter estimation.
4. Understand the theoretical basis for several commercial adaptive controllers.

PREREQUISITE MATHEMATICAL SKILLS

1. Laplace transforms.
2. Z-transforms.
3. Numerical integration of difference equations.

PREREQUISITE ENGINEERING AND SCIENCE SKILLS

1. Servo, regulatory control.
2. Block diagram analysis.
3. Transfer functions (s and z domain).
4. Controller tuning correlations.
5. Stability of closed-loop system.
6. Process reaction curve method.
7. Parameter estimation by linear regression.

Conventional feedback controllers applied to chemical processes inevitably require on-line tuning of the controller constants to achieve satisfactory performance. Often, chemical processes are nonlinear and process dynamics vary with changes in operating conditions. Thus, if a process exhibits time-varying dynamics and the operating conditions are changed, the controller may need to be re-tuned in order to provide satisfactory performance at the new operating condition.

In practice, controller constants are typically selected by trial and error experiments. In one approach the controller output signal is manually changed by the operator to introduce an upset and the control system response, i.e., the controlled variable, is then observed. Subsequently, the controller parameters are changed, and the procedure is repeated. The tuning process could take minutes, hours, or even days, depending upon the process response time for recovery after an upset.

Figure 1 shows a representative comparison of set point changes for the controlled variable before and after re-tuning, indicating the potential performance improvement which could be obtained by re-tuning. If the process, operating conditions, or environment changes significantly and quite frequently, then *adaptive control* techniques should be considered. An adaptive control system is one in which the controller parameters are adjusted automatically to compensate for changing process conditions. A variety of adaptive control techniques have been proposed for situations where the process changes are largely unknown and also for the more restrictive class of problems where the changes are known or can be anticipated.

Catalytic reaction systems are a notable example of an important process where unmeasurable changes occur. As Lee and Weekman (*1*) have noted:

> "Catalytic processes are notorious for changes in catalyst behavior, and adaptive loops are found quite commonly in such applications. The ad hoc development of an adaptive loop has probably saved more advanced control projects from failure than any other single technique."

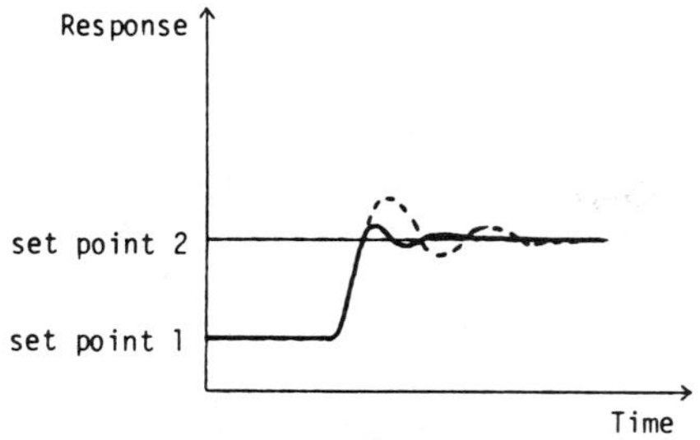

Figure 1. Closed-loop process response before re-tuning (dashed line) and after returning (solid line).

Other examples of changing process conditions which may require controller retuning or adaptive control are:

1. Heat exchanger fouling.
2. Frequent equipment changes (failures, start-up, shutdown).
3. Large, frequent disturbances (feed composition, fuel quality, etc.).
4. Ambient conditions (rain storms, 24-hour daily cycle, etc.).
5. Product specifications (i.e., "grade changes").

History of Adaptive Control

The early development of adaptive control strategies was motivated by the design of autopilots for high performance aircraft and rockets. Since the response characteristics of these aircraft varied significantly during flight conditions, classical linear controllers with constant controller parameters did not always provide satisfactory control. Consequently, considerable research activity during the 1950's was concerned with developing "self-adaptive" systems which would automatically adjust to changing flight conditions. These early efforts were largely unsuccessful and were characterized by "...a lot of enthusiasm, bad hardware and nonexisting theory" (*2*). During the 1960's, design strategies for adaptive control were placed on a more secure theoretical basis by the introduction of modern control concepts, especially those from stability theory. Renewed interest in adaptive control occurred in the 1970's due to significant theoretical developments such as self-tuning control systems and the widespread availability of inexpensive digital control hardware (e.g., microprocessor-based systems).

The wide variety of adaptive control strategies that have been proposed can be conveniently (but somewhat arbitrarily) placed into four categories:

1. Adaptive controllers designed using a quadratic cost function (e.g., self-tuning regulators and controllers).
2. Design methods based on stability theory (e.g., model reference adaptive control).
3. Pole assignment techniques.
4. Miscellaneous approaches, such as pattern recognition.

Seborg, et al. (*3*) have written an extensive review with over 200 references on the use of adaptive control techniques in the field of process control. In recent years, general purpose, adaptive control systems have appeared as commercial products in Europe, Japan and North America (*4*).

Adaptive Control Systems

It is convenient to distinguish between two general categories of adaptive control problems. The first category of adaptive control problems consists of situations where the process changes under consideration can be anticipated or measured. If the process is reasonably well understood, then it may be feasible to adjust the controller constants in a systematic fashion as process conditions change. For example, if a process is operated over a range of conditions, one approach would be to store a different set of controller constants for each operating condition. Alternatively, the controller constants could be expressed as continuous functions of particular process variables which produce the changes in process conditions. These strategies are examples of *programmed adaptation* (*5, 6*) since the effects of the process changes are predictable and thus can be programmed into the control system. In a sense, programmed adaptation is a feedforward type of strategy for adaptive control.

The second category consists of problems where the process changes cannot be measured nor predicted. In this difficult situation the adaptive control strategy must be implemented in a feedback manner since there is little opportunity for a feedforward type of strategy. This type of adaptive control is referred to as *self-tuning control.*

Programmed Adaptation

Liptak (*5*) has presented an example of a control problem where programmed adaptation has been proposed, namely a once-through boiler. Here feedwater passes through a series of tube sections before emerging as superheated steam, the temperature of which must be accurately controlled. The feedwater flow rate has significant effect on both the steady-state and dynamic behavior of the boiler. For example, Figure 2 shows typical open-loop responses to a step change in firing rate for two different feedwater flow rates, 50% and 100% of the maximum flow.

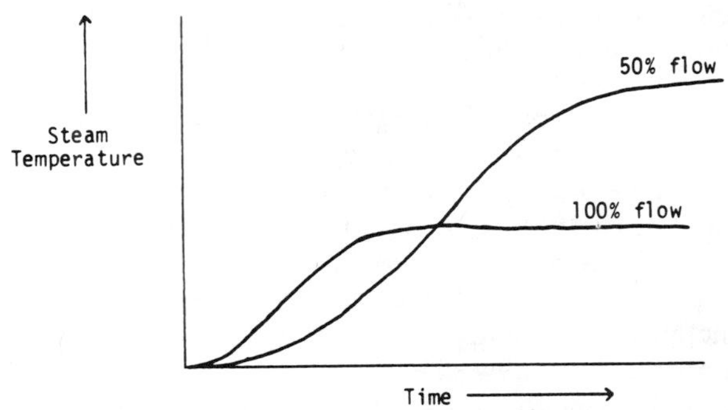

Figure 2. Open-loop step responses for a once-through boiler.

At 50% flow, the steady-state gain, time delay and dominant time constant are all about twice as large as the corresponding values for 100% flow. The solution (5) to this control problem is to make the PID controller settings vary with w, the fraction of full-scale (100%) flow, in the following manner:

$$K_c = wK_c^f \quad (1)$$
$$\tau_I = \tau_I^f / w$$
$$\tau_D = \tau_D^f / w$$

where K_c^f, τ_I^f, and τ_D^f are the controller constants for 100% flow. Note that this recommendation for programmed adaptation is qualitatively consistent with the rules for tuning controllers. Perhaps the most important parameter listed above is K_c, which needs to be adjusted so that the product of the process gain K_p and K_c is a constant, in order to maintain consistent stability margins. Shinskey (6) has noted that in most applications the process gain is quite susceptible to changes.

Stephanopoulos (7) has described another case where the controller gain needs to be adjusted on-line or adaptively. As is known from stability theory, it is desirable to keep the product of the control loop gain constant, i.e.,

$$K_{\text{loop}} = K_p K_m K_c K_v = \text{constant} \quad (2)$$

where K_m = measurement gain and K_v = control valve gain. If the valve gain varies in a nonlinear fashion with respect to stem position, and this variation is known, then we can program K_c to vary according to the relation,

$$K_c = \frac{K_{\text{loop}}}{K_p K_m K_v} \quad (3)$$

By knowing the valve position, K_v can be determined, and K_c is adjusted to keep K_{loop} constant (Equation 3).

In the once-through boiler example, step responses were available to categorize the process behavior for two different conditions. In other problems, dynamic response data are not available but some knowledge of process nonlinearities is. For pH control problems involving a strong acid and a strong base, the pH titration curve is very nonlinear with a process gain that can change by several orders of magnitude. Consequently, special nonlinear controllers, both adaptive and nonadaptive, have been developed for pH control problems (6).

For some types of adaptive control problems, the changes in steady-state and dynamic response characteristics can be related to the value of the controlled varible. For example, in a temperature control loop where the process gain varies with temperature, the controller gain could be made a function of the controlled variable, temperature. Controllers are now commercially available which allow the user to vary K_c as a piecewise linear function of the error signal, e, as shown in Figure 3. Intuitively,

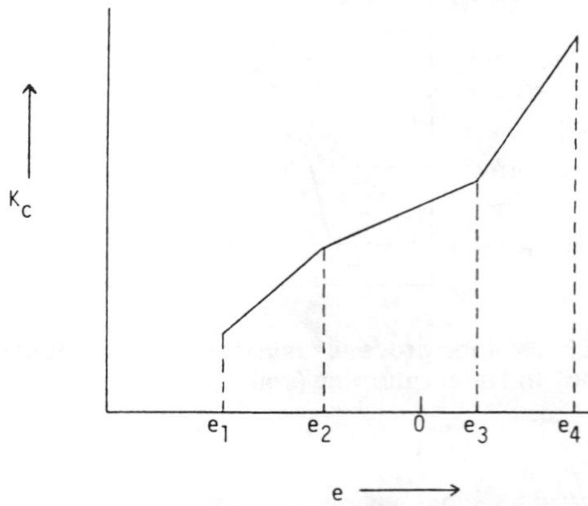

Figure 3. An adaptive controller where K_c varies with error signal, e.

if the process gain K_p varies in a known manner, one should vary K_c so that the product, K_pK_c, is constant. This strategy, nonlinear gain compensation, would tend to keep the open-loop gain constant and thus maintain a specified margin of stability.

Self-tuning Control

If process changes cannot be measured nor anticipated, then programmed adaptation cannot be used. An alternative approach would be to update the parameters in a process model on-line and then base the control calculations on the updated model. For example, the controller constants could be expressed as a function of the parameters in the process model. The estimates of these parameters could be updated on-line as process data are received. This approach is known as "self-tuning" or "self-adaptive" control.

Two configurations can be used to implement self-tuning control, as shown in Figures 4 and 5. In Figure 4, three stages are involved: the estimation of the process model parameters based on measuring the input and output variables, calculation of the controller parameter settings, and implementation of the resulting control action. The second approach, model-reference adaptive control (3, 8), is shown in Figure 5. It employs a reference model to generate the desired response of the controlled output C to a set point change R. Any deviation from the reference model (ΔC

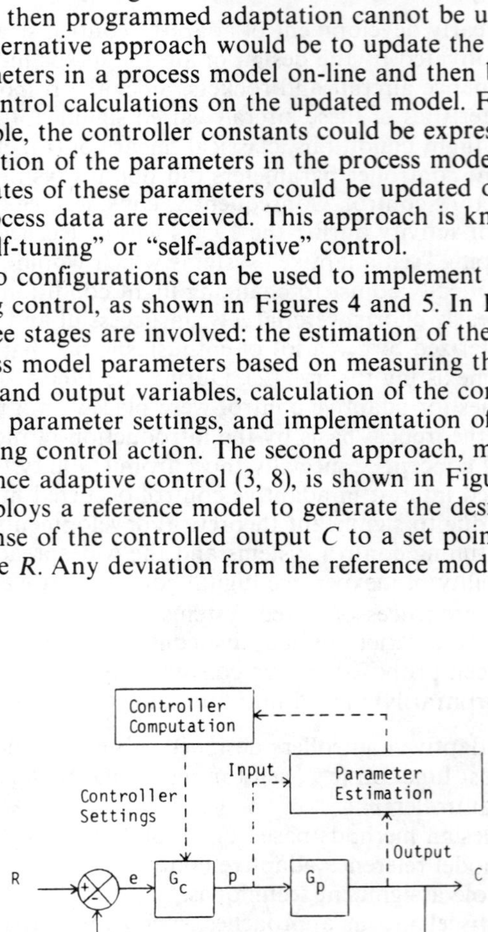

Figure 4. Block diagram for self-tuning control.

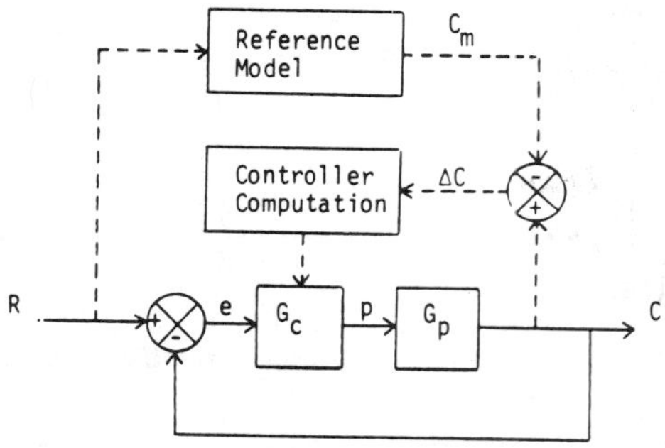

Figure 5. Block diagram for model reference adaptive control.

in Figure 5) is used to update the controller settings. The updating strategy is generally selected to ensure that ΔC is reduced to zero in a reasonable amount of time (*3*).

Most real-time parameter estimation techniques require that an external forcing signal be deliberately introduced to allow process identification (*9, 10*). Such an input signal can be introduced through the set point or superimposed on the controller output. The input can be a step change, a sinusoidal variation, white noise, or a sequence of random rectangular pulses (referred as a pseudo random binary sequence, or PRBS). While the latter two approaches have some statistical advantages for parameter estimation, they may not excite the process sufficiently and may be obscured by process noise. The model parameters can be estimated using a variety of techniques, such as least squares, cross-correlation analysis, or Fourier analysis (*9, 10, 11*).

Perhaps the simplest case of self-tuning control involves periodic step-testing of the process, e.g., employing the so-called process reaction curve. Suppose a first-order plus time-delay model is to be fitted to the response data. This can be done by graphical analysis to estimate the process gain K_p, time constant τ, and time delay, θ (*12*). Then one can employ the Cohen-Coon correlations to specify the controller parameters. For a PID controller, the Cohen-Coon settings are

$$K_c = \frac{1}{K_p}\frac{\tau}{\theta}\left(\frac{4}{3} + \frac{\theta}{4\tau}\right) \tag{4}$$

$$\tau_I = \theta \cdot \frac{32 + 6\theta/\tau}{13 + 8\theta/\tau} \tag{5}$$

$$\tau_D = \theta \cdot \frac{4}{11 + 2\theta/\tau} \tag{6}$$

Other tuning methods such as correlations based on integral absolute error could also be used (*12*).

Commercial Controllers

In the section on history of adaptive control, we identified four categories of adaptive control strategies. Specific techniques in all four categories have been tested experimentally (*3*). Several commercial adaptive controllers have been developed, arising out of categories 1, 3, and 4. While design methods based on stability theory (category 2) have been used on a pilot-plant level, commercial controllers of this type are currently not available. Three commercial controllers will be discussed in more detail:

1. Leeds and Northrup (pole assignment).
2. Self-tuning regulator (quadratic cost function).
3. Foxboro (pattern recognition).

All three adaptive controllers are microprocessor-based and hence the control algorithms are formulated in discrete-time rather than continuous time.

Leeds and Northrup

Leeds and Northrup (L&N) in 1981 announced the availability of a self-tuning PID controller, (*4, 13*). This controller provides the capability of operating as a conventional controller, where the operator enters the settings, or the controller can automatically tune itself in response to process changes. In general, the self-tuning operation is designed to yield an exponential approach of the process output to a step change in set point. So far this controller has had its main application in fuel-fired or electrically-heated processes. In these processes, process changes which might require re-tuning of the controller include:

1. Mechanical differences, such as primary element location and location of the load.
2. Energy variations such as Btu content or firing rate.
3. Air supply variation, such as temperature or flue restrictions and seasonal ambient temperature changes.

During the self-tuning procedure, a controlled upset—preferably as large as the system will handle—is introduced into the system in order to determine process time constants. The controller then observes the system response and necessary corrections are made to the original set of constants. This parameter estimation step is then followed by calculation of the controller settings.

Andreiev (*13*) and Hoopes et al. (*4*) have described the tuning procedure for the Leeds and Northrup (L&N) controller. The self-tuning controller can operate in four different modes:

1. Identification mode.
2. Examination mode.
3. Fixed mode.
4. Self-tuning mode.

The first step in automatic tuning requires operator interaction with the controller during the identification mode. The operator enters (by means of a keyboard) an acceptable set point change—similar to the one used in manual, trial and error tuning—and the desired closed-loop response time. The adaptive controller then automatically introduces the set-point upsets. By monitoring the actual process response, it forms an internal model of the process. When the model matches the actual process, the computed controller constants appropriate to the process characteristics may be displayed in preparation for the next step. The controller tuning constants are designed for an exponential approach to set point, with no overshoot (overdamped response).

In the examination mode, the operator may view the computed controller constants and choose to operate the process with the controller in either the fixed or the self-tuning mode. In the fixed mode, the adaptive controller uses the previously determined PID settings but continuously updates the process model and calculates the corresponding controller constants. The operator has the option to enter these new constants at any time, thus bypassing the identification mode. In the self-tuning mode, the computed constants will be continuously updated by the controller to adapt to changing process conditions and/or loads.

To initiate the self-tuning mode, the operator must specify the desired response time and the magnitude of the step change of set point that is used to identify the process characteristics. The response time is defined as the number of minutes it takes the process in a closed-loop system to complete 90% of a step change in set point. Figure 6 shows typical responses for set-point changes using a variable response time.

A typical test sequence for the identification mode is shown in Figure 7 (*13*). At t_1, the controller was placed in the identification mode. At t_2 and t_3, predetermined set-point changes were implemented. By monitoring this response, the self-tuning controller developed a process model and computed the associated PID controller constants; it then indicated that the identification mode was completed. At t_4, the operator pressed the ENTER button and simultaneously transferred the newly computed PID settings into the controller and initiated the self-tuning mode.

At t_5, the set point was increased and the process

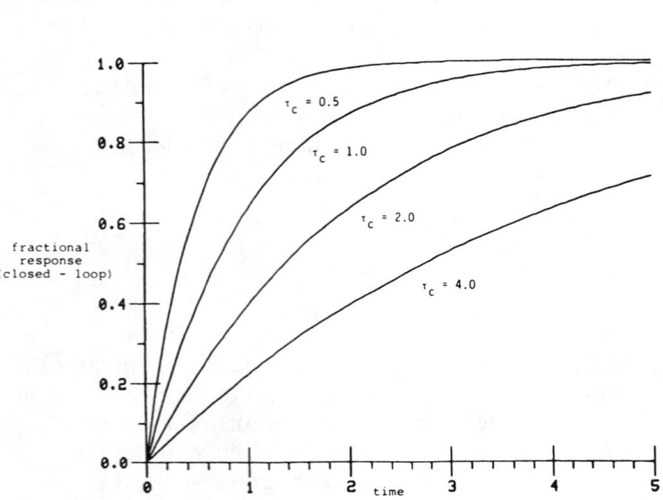

Figure 6. Closed-loop responses for exponential approach to set point (τ_c = time constant of response).

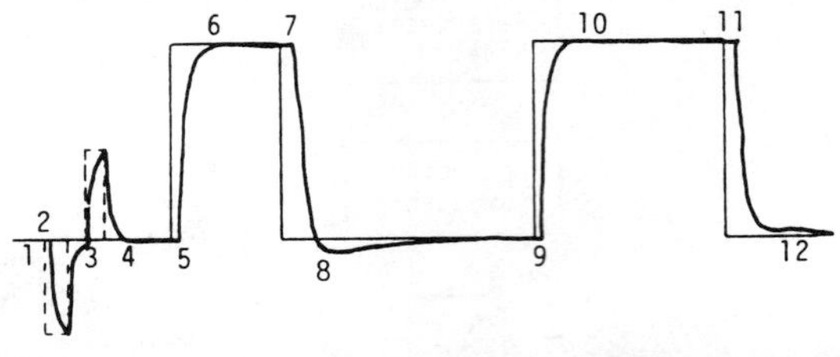

Figure 7. Test sequence for model identification (Leeds and Northrup controller).

variable responded smoothly. At t_6, the process response time was increased to accommodate a required process change (heating rate in this case), and at t_7 the set point was reduced. However, at t_8, the process variable overshot the new set point because of the changed process conditions between t_7 and t_9. The self-tuning function was updating the process model and changing the computed PID tuning constants. Consequently, when the set point was increased at t_9, and later decreased at t_{11}, the process variable responded smoothly with no overshoot, reaching the set points at t_{10} and t_{12}, respectively.

There are some additional features of the controller. For example, the controller will recognize a power loss during the self-tuning mode of operation and will automatically switch into a safe mode of operation to protect the process. When power is restored, the controller will return the process, again automatically, to the previous set point before switching into the self-tuning mode.

Theoretical Basis for the Controller

The self-tuning procedure is based on the assumption that the process model is second order plus time delay and the desired response of the closed-loop system is first order (exponential rise to a positive step change in set point). In continuous time, the closed-loop transfer function is

$$\frac{C}{R} = \frac{G_c G_p}{1 + G_c G_p} = \frac{e^{-\theta s}}{1 + \tau_c s} \qquad (7)$$

where, G_c is the controller transfer function, τ_c is the time constant of the closed-loop response, and G_p is the second-order process model. If the time delay (θ) is zero,

$$G_p(s) = \frac{K_p}{(1 + \tau_1 s)(1 + \tau_2 s)} \qquad (8)$$

Solving Equation 7 for G_c, using Equation 8 for G_p,

$$G_c(s) = \frac{(1 + \tau_1 s)(1 + \tau_2 s)}{K_p \tau_c s} \qquad (9)$$

or

$$G_c(s) = \frac{1}{K_p \tau_c}\left[(\tau_1 + \tau_2) + \frac{1}{s} + \tau_1 \tau_2 s\right] \qquad (10)$$

Note that this controller is PID, with the individual mode settings determined as an explicit function of the model parameters (K_p, τ_1, τ_2) and the response time, τ_c.

$$K_c = \frac{\tau_1 + \tau_2}{K_p \tau_c} \tag{11}$$

$$\tau_I = \tau_1 + \tau_2 \tag{12}$$

$$\tau_D = \frac{\tau_1 \tau_2}{\tau_1 + \tau_2} \tag{13}$$

The L&N controller is actually based on a discrete-time formulation, which is readily implemented on a digital microcomputer, and is adjusted to account for a known time delay. The second-order process transfer function is denoted in discrete time as

$$HG(z) = \frac{(b_1 z^{-1} + b_2 z^{-2}) z^{-N}}{1 + a_1 z^{-1} + a_2 z^{-2}} \tag{14}$$

where z is the z-transform variable, N is the time-delay (nearest integer value of $\theta/\Delta t$, where Δt is sampling period) and a_1, a_2, b_1, and b_2 can be related to the process gain and time constants of the second-order continuous model as well as to the sampling period (*12*). The parameters a_1, a_2, b_1 and b_2 are estimated on-line while the controller is operating in the feedback mode. The time delay N must be specified by the operator.

To obtain a PID digital controller, Hoopes et al. (*4*) assume a closed-loop response formula,

$$\frac{C}{R} = \frac{(1 - e^{-\Delta t/\tau_c}) z^{-N-1}}{1 - e^{-\Delta t/\tau_c} z^{-1}} \tag{15}$$

By including the time-delay term in C/R, time-delay compensation is automatically provided. The resulting transfer function for the digital controller is

$$D(z) = \frac{(1 - A)}{1 - Az^{-1} - (1 - A) z^{-N-1}} \cdot \frac{1 + a_1 z^{-1} + a_2 z^{-2}}{b_1 + b_2 z^{-1}} \tag{16}$$

where A is the tuning parameter, $e^{-\Delta t/\tau_c}$, corresponding to the response time. This is commonly referred to in the digital control literature as Dahlin's controller (*12*, *14*). For this type of controller, the closed-loop response time remains constant, although the controller settings vary when the process model changes (Equation 16).

There are several potential deficiencies in the L&N controller algorithm, most of which stem from the use of Dahlin's controller. Problems with ringing and time-delay changes could cause this control algorithm to have poor performance. In order to reduce ringing, the term $(b_1 + b_2 z^{-1})$ in Equation 16 could be replaced by $(b_1 + b_2)$ (*14*). An alternative algorithm developed by Vogel and Edgar (*15*) overcomes these difficulties and thus would be preferred; the Vogel-Edgar technique has been tested successfully on a heat exchanger loop and a pilot-scale distillation column.

Model Identification

The model parameters (a_1, a_2, b_1, and b_2) must be updated by the series of step changes described earlier. No change in delay time N is considered. For $N = 0$, the process model given in Equation 14 corresponds to the difference equation,

$$y_n = -a_1 y_{n-1} - a_2 y_{n-2} + b_1 p_{n-1} + b_2 p_{n-2} \tag{17}$$

where a_1, a_2, b_1, and b_2 are parameters to be fitted, y_n is the controlled variable, and p_n is the controller output at the nth time step. Note that the model parameters appear linearly in the model, thus the best fit can be obtained by application of least squares estimation theory (*10*). This yields a set of four linear equations in four unknowns (a_1, a_2, b_1, and b_2) which can be readily solved on a digital computer (see Module A3.6 for the procedure). However, in adaptive control it is preferable to use on-line or sequential estimation, which is tantamount to iterative improvement of the estimates of the unknown parameters at each time step. Young (*16*) has described the calculations involved, which are readily implemented on a microcomputer. The specific strategy used in the L&N controller is based on the "instrumental variable" method, as applied by Touchstone and Corripio (*17*). The interested reader is referred to their paper as well as to Reference (*16*) for more details on recursive estimation algorithms.

Self-tuning Regulator

The self-tuning regulator (STR) as formulated by Åström (*11*) is based on a so-called "minimum variance control law." In Equation 17, if previous values of y and p are stored and the model parameters are known, y_n can be predicted. In order to keep the value of y_n at the set point (R), Equation 17 can be solved to obtain the appropriate value of the controller output,

$$p_{n-1} = \frac{1}{b_1}(R + a_1 y_{n-1} + a_2 y_{n-2} - b_2 p_{n-2}) \tag{18}$$

Therefore, as long as the model parameters are known, the controller output can be adjusted accordingly for the next time step. This is called a "one-step ahead" control algorithm.

The STR is implemented by periodic updating of the model parameters using sequential or recursive estimation (*16*), followed by application of Equation 18. At each sampling instant,

1. The values of p_i and y_i would be recorded during the transient phase of the operation.
2. Using recursive estimation, new parameter estimates are obtained.
3. A new value of p_{n-1} is computed using Equation 18.

A disadvantage of the one-step ahead control algorithm, Equation 18, is that no tuning parameter other than the sampling time is available. This often leads to an overactive controller which can saturate the control valve. Clarke and Gawthrop have developed a modified STR which avoids constraint violation (*18*).

The above discussion presents a very simplistic view of the STR; more sophisticated versions have been implemented commercially by several companies, including ASEA in Sweden and AccuRay in the U.S.A. (*3*). The principal applications areas have been in paper machine control; a list of applications papers (e.g., cement mills, heat exchangers, pH control) has been given by Åström (*2*) and Seborg et al. (*3*). Improvements in the STR are currently under intensive research (*3*).

Pattern Recognition

This controller has been in development at Foxboro since 1970, and some details of its operation have been presented by Bristol (*19 - 21*). This controller evaluates the closed-loop response to a step change in set point. Figure 8 shows a typical set-point response and a number of characteristic, measurable parameters associated with the transient behavior. No explicit identification of model parameters such as time constants is carried out in this approach. The controller equation proposed by Bristol (*19*) has the same form as a velocity-type PI algorithm:

$$\Delta p_n = \alpha_1 \Delta e_n + \alpha_2 e_n \qquad (19)$$

where Δp is the change in controller output, e_n is the error, and α_1 and α_2 depend on e_n and the transient characteristics illustrated in Figure 8. In addition, there are several tuning parameters included in α_1 and α_2, which must be determined by the user. An updating strategy for α_1 and α_2 has been discussed by Bristol. Our discussion of the Foxboro controller here is necessarily limited because a detailed description of the control strategy is not currently available in published form, presumably for proprietary reasons.

SUMMARY

This module has been designed to provide a summary of recent developments of commercial adaptive controllers for chemical process control; in so doing, we have only briefly addressed the large body of research literature dealing with adaptive control strategies. Four classes of adaptive control strategies have been identified, and the role of parameter estimation in adaptive control has been discussed. The theoretical bases for three commercial controllers have been presented, along with their advantages and disadvantages.

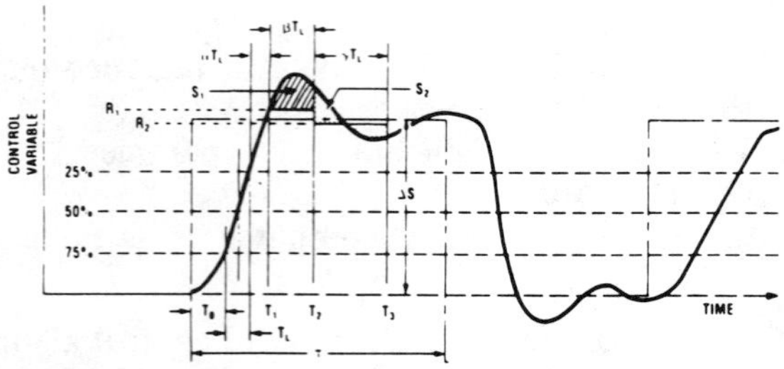

Figure 8. The transient response of the process as identified by the adaptation scheme.

NOMENCLATURE

A	=	tuning parameter in Dahlin's controller
a_i, b_i	=	discrete-time model parameters
C, c	=	controlled variable
e	=	error = set point minus measurement
G_c	=	controller transfer function
G_p	=	process transfer function
HG	=	discrete process model (with zero-order hold)
K_c	=	controller gain
K_{loop}	=	loop gain
K_m	=	measurement gain
K_p	=	process gain
K_v	=	control valve gain
N	=	integer value of time delay
p	=	controller output
R	=	set point
s	=	Laplace transform variable
w	=	fraction of full-scale flow
y	=	controlled variable
z	=	z-transform variable
α	=	controller parameter
θ	=	time delay or dead time
Δt	=	sampling period
τ	=	process time constant
τ_c	=	response time (closed-loop)
τ_D	=	derivative time
τ_I	=	integral time

LITERATURE CITED

1. Lee, W. and V. M. Weekman, Jr., "Advanced Control Practice in the Chemical Process Industry: A View From Industry," *AIChE J.*, *22*, 27 (1976).
2. Åström, K. J., "Theory and Applications of Adaptive Control—A Survey," *Automatica*, *19*(5), 471 (1983).
3. Seborg, D. E., Shah, S. L. and T. F. Edgar, "Adaptive Control Strategies for Process Control: A Survey," AIChE Diamond Jubilee Meeting, Washington, D.C., October, 1983.
4. Hoopes, H. S., W. M. Hawk, and R. C. Lewis, "A Self-tuning Controller," *ISA Trans.*, *22*, (3), 49 (1983).
5. Liptak, B. G. "Instrument Engineers Handbook," Vol. II, Chilton Book Company, Philadelphia, PA (1970).
6. Shinskey, F. G., "Process Control Systems," 2nd Edition, McGraw-Hill, New York (1979).
7. Stephanopoulos, G., "Chemical Process Control," Prentice-Hall, Englewood Cliffs, NJ (1983).
8. Landau, Y. D., "Adaptive Control," Marcel Dekker, New York (1978).
9. Isermann, R., "Digital Control Systems," Springer-Verlag, Berlin (1981).
10. Eykhoff, P., "System Identification," Wiley, London (1974).
11. Åström, K. J., "Theory and Application of Self-Tuning Regulators," *Automatica, 13*, 457 (1977).
12. Smith, C. L., "Digital Process Control," In Text, Scranton, PA (1972).
13. Andreiev, N., "A New Dimension: a Self-Tuning Controller that Continually Optimizes PID Constants," *Control Eng.*, 84, August (1981).
14. Corripio, A. B., "Digital Control Techniques," Module 3.5 in AIChEMI Series, Series A, Vol. 3 (Process Control) (1983).

15. Vogel, E. F., and T. F. Edgar, "Application of an Adaptive Pole-Zero Placement Controller to Chemical Processes with Variable Dead Time," *Proc. Amer. Cont. Conf.*, 536, Arlington, VA (1982).
16. Young, P. C., "Applying Parameter Estimation to Dynamic Systems—I, II," *Control Eng.*, *16*(10), 119 (1969).
17. Touchstone, A. T., and A. B. Corripio, "Adaptive Control Through Instrumental Variable Estimation of Model Parameters," Instrumentation in Chemical and Petroleum Industries, ISA, Anaheim, CA (1977).
18. Clarke, D. W., and P. J. Gawthrop, "Implementation and Application of Microprocessor-Based Self-Tuners," *Automatica*, *17*, 233 (1981).
19. Bristol, E. H., Inaloglu, G. F. and J. F. Steadman, "Adaptive Process Control by Pattern Recognition," *Inst. Cont. Syst.*, *43*(3), 101, March (1970).
20. Bristol, E. H., "Pattern Recognition: An Alternative to Parameter Identification in Adaptive Control," *Automatica*, *13*, 197 (1977).
21. Bristol, E. H., "The Design of Industrially Useful Adaptive Controllers," *ISA Trans.*, *22*(3), 17 (1983).

STUDY PROBLEMS

1. What process characteristics are conducive to use of adaptive control? Suggest some possible applications in the process industries.

2. Give an example of programmed adaptation.

3. Describe the necessary computations for "self-tuning" control.

4. How can continuous cycling of a process at its stability limit be used in conjunction with adaptive control? Would this approach be advantageous compared to other adaptive controllers?

5. Contrast the operation of the Leeds & Northrup adaptive controller and the self-tuning regulator (STR).

6. What conclusions can be drawn if overshoot occurs in the Leeds and Northrup controller for a set point change?

7. Name some undesirable features of currently available commercial adaptive controllers.

HOMEWORK PROBLEMS

1. You are given the following second-order pulse transfer function:

$$HG(z)=\frac{0.02797\,z^{-1}+0.02342\,z^{-2}}{1-1.535\,z^{-1}+0.5866\,z^{-2}}$$

This model corresponds to a second-order continuous model with a zero-order hold and $K_p=1$, $\tau_1=3$, $\tau_2=5$, and $\theta=0$ sampled at $\Delta t=1$.

a) Using $\tau_c=1$, derive Dahlin's controller (ringing and nonringing) and simulate the closed-loop response for a set point change.
b) Increase the model gain by increasing the first numerator coefficient from $b_1=0.02797$ to $b_1=0.035$. Using the nonringing controller in (a), perform the same set point change. Compare the response with that in (a). These results will show how the response deteriorates when the controller does not adapt to the new model.
c) Would decreasing the controller gain in Dahlin's controller such that $K_p K_c=$ constant be a satisfactory way to treat adaptively the increased model gain? Why or why not?
d) Repeat the exercise in (a) and (b) for one-step ahead control, such as that used in the self-tuning regulator.

Module A4.6

Dead Time Compensation

Ram Lavie

Department of Chemical Engineering,
Technion-Israel Institute of Technology
Haifa, Israel

OBJECTIVES

At the completion of this module, the student should understand the undesirable effects of dead time in a feedback control loop and how stability can be promoted by using dead time compensation. The student will:

a. Identify the causes of dead time and formulate the control problem in standard block diagram form, isolating those elements that must enter the compensator.
b. Evaluate the performance of the compensated system as compared to the noncompensated original problem.
c. Consider some practical limitations in the implementation of dead time compensation.

PREREQUISITE MATHEMATICAL SKILLS

1. Laplace Transforms.

PREREQUISITE ENGINEERING AND SCIENCE SKILLS

1. Linear Feedback Control Analysis and Synthesis Methods.

INTRODUCTION

Feedback control, while being a most effective and widely used control strategy, is seriously limited in its ability to deal with processes that include substantial time delays.

Dead time is a delay described mathematically by:

$$y(t)=x(t-\theta) \tag{1}$$

or

$$\frac{Y(s)}{X(s)}=e^{-\theta s} \tag{2}$$

Typically, it appears in a process model as the result of:

a) Transportation lag in long pipes or other process equipment.
b) Delayed availability of measured quantities, (typically, results of quantitative chemical analysis).
c) Delays in signal transmission and in activation of control devices.
d) Intentional delays in the activation of devices because of operational or safety considerations.
e) Time necessary for the controller (an automatic device or a human operator) to translate a state deviation into a control decision.
f) Approximation of models describing high order processes by low order dynamics plus dead time.

Because of its sudden nature, a dead time process output does not provide any "early warning" indication about a disturbance entering the process.

Statement of the Problem

The presence of dead time in a process that is controlled by feedback is determined in three ways: (See Figure 1).

a) A disturbance $D(s)$ entering the process is detected at the process output only after a while θ_D, thus delaying the application of corrective action. This situation can only be prevented by detecting the disturbance at its source, predicting its future effects and counteracting them before they appear at the process output. This is the realm of feedforward control and it will not be discussed here.
b) A change in set point $R(s)$ to the feedback controller will similarly start taking effect at the process

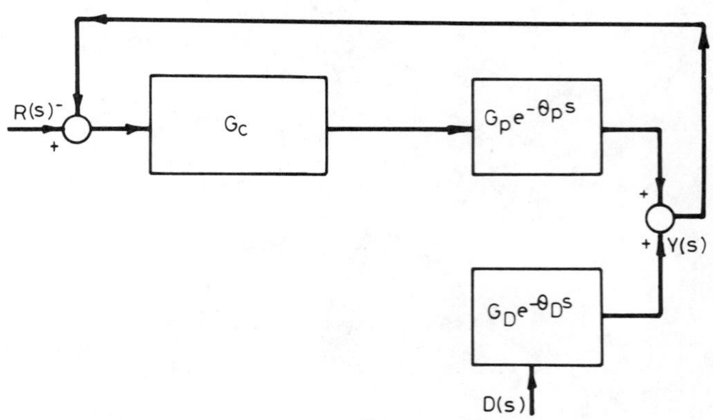

Figure 1. Conventional feedback control loop.

output only after a time delay θ_p. This is unavoidable and must be taken into account when taking supervisory decisions.

c) Every control decision corrects a deviation that originated some while back in time. It may thus be irrelevant to the necessary correction at that particular time. Uncoordinated control tends to be unstable. The purpose of dead time compensation is to overcome this undesirable tendency.

Figure 1 is a block diagram describing conventional feedback control of a process that includes dead time. G_c, G_p, G_D are transfer functions that do not include dead time. The process transfer function $[G_p(s \cdot e^{-\theta_p s}]$ includes, in addition to the process itself, everything in the loop (measurement, transmission, control valve) other than the controller G_c itself.

The dynamic properties of the control loop are governed by the solutions to the characteristic equation:

$$1+G_c(s)G_p(s)e^{-\theta_p s}=0 \tag{3}$$

Using a first–order Pade approximation for the dead time,

$$e^{-\theta_p s}=\frac{(s-2/\theta_p)}{(s+2/\theta_p)} \tag{4}$$

Evidently, the real positive zero at $s=2/\theta_p$ must eventually, for some controller gain, cause instability.

Compensation

Suppose $G_p(s)$ and θ_p are known accurately. Then it may be possible to construct a dynamic simulator described by the transfer function:

$$G_m(s)=(1-e^{-\theta_p s})\cdot G_p(s) \tag{5}$$

and install it on a feedback path around the controller G_c, as described in Figure 2.

The element $G_m(s)$ is called a *dead time compensator*. The element $(1-e^{-\theta_p s})$ is sometimes called a *pulse generator*. The addition of the compensator is equivalent to the replacement of the controller G_c in Figure 1 by

$$G_{cm}(s)=\frac{G_c(s)}{1+(1-e^{-\theta_p s})G_c(s)\cdot G_p(s)} \tag{6}$$

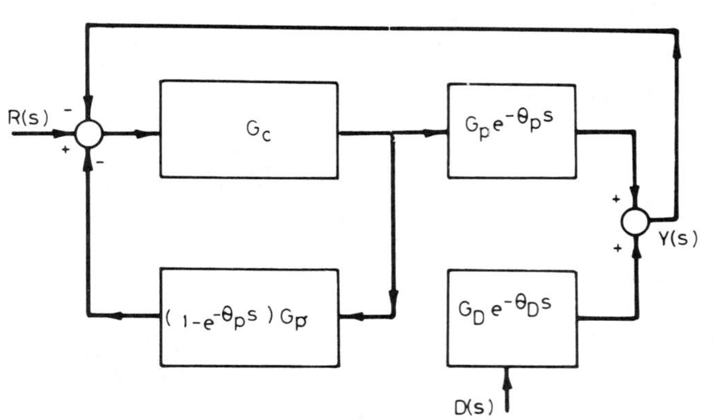

Figure 2. Compensated feedback control loop.

a form including process model components $G_p(s)$ and $e^{-\theta_p s}$. The closed loop is now described by:

$$Y(s)=\frac{1}{1+G_{cm}G_p e^{-\theta_p s}}[G_{cm}G_p e^{-\theta_p s}\cdot R(s) + G_D e^{-\theta_D s}\cdot D(s)] \tag{7}$$

Inserting Equation 6 into Equation 7,

$$Y(s)=\frac{1}{1+G_cG_p}[G_cG_p e^{-\theta_p s}\cdot R(s) + \frac{G_cG_D e^{-\theta_D s}}{G_{cm}}\cdot D(s)] \tag{8}$$

is obtained.

Dead time (and its undesirable dynamic effect) has thus been eliminated from the characteristic Equation 3.

Note that dead time still appears in the nominator of the transfer functions. Thus, while the dynamics of the control loop have been improved by the compensation, a delay between a change in set point $R(s)$ or a disturbance $D(s)$, and the measured response $Y(s)$ of the process still remains.

Limitations

The method has several limitations:

a) A reasonably accurate model of the process, including all dead times, is essential to the application of dead time compensation.

b) Means must be available to simulate the process and the dead time as part of the controller. Implementation of the method waited several years because a flexible simulator for dead time was not generally available at that time.

c) The compensator can eliminate the dynamic consequences of only such dead times that are external to the controller G_c. If, for example, it would take the computing device G_c a significant amount of time to generate a control decision, then the present method would be unable to compensate for that delay. Fortunately, this is seldom the case.

Figure 3 is a computer program which demonstrates the performance of deadtime-compensated feedback control for the control of a first–order process with dead time.

The user specifies in an initial dialogue some process and PI controller parameters, which are known to yield an unstable system in the presence of dead time. The program then simulates the compensated controlled system and prints out the process response to user-initiated changes in set point or load. Table 1 is a sample output of this program, named DTC.

Program Name: *DTC*
Class: *Limited*
Type: *Self-Standing*

Title: Dead Time Compensation.
Author: Gaddi

Program Development:

1. Computer used: PDP 8e
2. Minimum Core: 16k
3. Operating Environment: Real Time (OS/8)
4. Language: R/T BASIC
5. Typical Run Time: One minute
6. Number of Source Cards: 174

PROGRAM LISTING

```
10  REM   NAME: XDTC
20  DIM C2(1000),C4(1000)
30  PRINT PNT(29);PNT(31)
40  PRINT "        D E A D    T I M E   C O M P E N S A T I O N"
50  PRINT "        ************************************************* "
60  PRINT
70  PRINT
80  PRINT "PROCESS CHARACTERISTICS"
90  PRINT "-----------------------"
100 PRINT
110 PRINT "PROCESS TRANSFER FUNCTION:  K1/(AS+1)"
120 PRINT "INSERT K1 ";
130 INPUT K1
140 PRINT "INSERT A ";
150 INPUT A
160 PRINT
170 PRINT
180 PRINT "PL CONTROLLER: KC*(1+1/TI*S) "
190 PRINT "INSERT KC ";
200 INPUT K
210 PRINT "INSERT TI ";
220 INPUT T2
230 PRINT
240 PRINT
250 PRINT "INSERT DEAD TIME (TAU) ";
260 INPUT T1
270 T3=T1/1000
280 T3=INT(T3*10**4+.5)/10**4
290 PRINT
300 PRINT
310 PRINT "INTEGRATION CHARACTERISITCS"
320 PRINT "---------------------------"
330 PRINT
340 PRINT "METHOD:  EULER INTEGRATION "
350 PRINT
360 PRINT "CHOOSE INTEGRATION STEP ( SHOULD BE > = ";T3;" ) ";
370 INPUT D
380 D1=INT(D*10**4+.5)/10**4
390 IF D1>=T3 THEN 420
400 PRINT "INTEGRATION STEP TOO SMALL. ENTER A GREATER ONE ! "
410 GO TO 370
420 PRINT "SET A TIME LIMIT FOR THIS RUN ";
430 INPUT F
440 PRINT "SELECT PRINTING INTERVAL ( SHOULD BE > = ";D1;" ) ";
450 INPUT S
460 S7=INT(S*10**4+.5)/10**4
470 IF S7>=D1 THEN 500
480 PRINT "PRINTING INTERVAL < INTEGRATION STEP : ENTER ANOTHER ONE !"
490 GO TO 450
500 PRINT
510 PRINT
520 PRINT
530 J9=0
540 S1=INT(S/D+.5)
550 I1=INT(T1/D+.5)
560 I2=INT(2*I1+.5)
570 PRINT "NOW CHOOSE R AND U, THE SET-POINT CHANGE AND TEH LOADL CHANGE,"
580 PRINT  "BY INSERTING THE MAGNITUDES OF THE STEP CHANGES YOU DESIRE."
590 PRINT  "WHAT STEP CHANGE DO YOU WANT IN THE SET-POINT ";
600 INPUT R
610 PRINT "AND IN THE LOAD ";
620 INPUT U
630 PRINT PNT(29);PNT(31)
640 FILEV#3:"TTY:"
650 GOSUP 1470
660 PRINT
670 PRINT
680 PRINT "DO YOU WANT TO CHANGE ANY DATA (ANSWER 'Y' OR "N") ";
690 INPUT AS
700 IF AS<>"Y" THEN 730
710 PRINT PNT(29);PNT(31)
720 GO TO 80
730 PRINT
740 PRINT "DO YOU WANT A PRINTOUT OF YOUR DATA (ANSWER 'Y' OR 'N') ";
750 INPUT DS
760 IF DS<>"Y" THEN 790
770 FILEV#3:"TT:"
780 GOSUB 1470
790 PRINT
800 PRINT "DO YOU WANT YOUR RESULTS PRINTED ON THE SCREEN OR AT THE  ";
810 PRINT "TELETYPE (ANSWER 'S' FOR SCREEN, 'T' FOR TELETYPE)";
820 INPUT RS
830 IF RS<>"S" THEN 860
840 PRINT PNT(29);PNT(31)
850 FOSUB 1670
860 IF RS<>"T" THEN 940
870 FILEV#3:"TT:"
880 PRINT#3:"TIME","  C    ","     C1        ","    C1 - C"
890 PRINT#3:"     ","PLANT  ","COMPENSATED","COMPENSATOR"
900 PRINT#3:"     ","OUTPUT","     OUTPUT  ","   EFFECT"
910 PRINT#3:"----------------------------------------------------------"
920 PRINT#3:
930 CLOSE#3
940 FOR T=D TO F  STEP  D
950 T=INT(T*10**4+.5)/10**4
960 J=J+1
970 I=I+1
980 E=R-C1
990 Z1=K*E
1000 Z=Z+Z1*D
1010 M=Z1+Z/T2
1020 M1=M+U
1030 W1+M/A-W/A
1040 W=W+W1*D
1050 V1+M1/A-V/A
1060 V=V+V1*D
1070 IF T>T1 GO  TO  1250
1080 C2(I)=K1*W
1090 C3=0
1100 C5=C2(I)
1110 C4(I)=K1*V
1120 C=0
1130 C1=C+C5
1140 B=C1-C
1150 IF INT(J+.5)<>S1  GO  TO  1430
1160 IF RS<>"S" GO  TO  1190
1170 J9=J9+1
1180 PRINT T,C,C1,B
1190 IF RS<>"T" THEN 1230
1200 FILEV#3:"TT:"
1210 PRINT#3:T,C,C1,B
1220 CLOSE#3
1230 J=0
1240 GO  TO 1430
1250 C3=C2(I-I1)
1260 C2(I-I1)=K1*W
1270 C5=C2(I-I1)-C3
1280 C=C4(I-I1)
1290 C4(I-I1)=K1*V
1300 C1=C+C5
1310 B=C1-C
1320 IF INT(J+.5) <>S1 GO  TO  1410
1330 IF RS<>"S" THEN 1360
1340 J9=J9+1
1350 PRINT T,C,C1,B
1360 IF RS<>"T" THEN 1400
1370 FILEV#3:"TT:"
1380 PRINT#3: T,C,C1,B
1390 CLOSE#3
1400 J=0
1410 IF I<>I2  GO  TO  1430
1420 I=I1
1430 IF RS<>"S" THEN 1450
1440 GOSUB 1640
1450 NEXT T
1460 STOP
1470 PRINT#3:"     D A T A    S U M M A R Y"
1480 PRINT#3:"     ------------------------ "
1490 PRINT#3:
1500 PRINT#3:"PROCESS TRANSFER FUNCTION: ";K1;"/ ( ";A;"* S + 1 )"
1510 PRINT#3:"PI CONTROLLER:  ";K;"* ( 1 + 1 / ";T2;"* S )"
1520 PRINT#3:"DEAD TIME:  TAU = ";T1
1530 PRINT#3:
1540 PRINT#3:
1550 PRINT#3:"SET-POINT  STEP CHANGE : ";R
1560 PRINT#3:"LOAD  STEP CHANGE : ";U
1570 PRINT#3:
1580 PRINT#3:
1590 PRINT#3:"INTEGRATION STEP  : ";D1
1600 PRINT#3:"FINISHING TIME    : ";F
1610 PRINT#3:"PRINTING INTERVAL : ";S7
1620 CLOSE#3
1630 RETURN
1640 IF INT(J9/13)<>J9/13 THEN 1730
1650 PRINT
1660 PRINT
1670 PRINT  "TIME","  C  ","     C1     ","   C1 - C"
1680 PRINT  "     ","PLANT","COMPENSATED","COMPENSATOR"
1690 PRINT  "     ","OUTPUT","  OUTPUT  ","   EFFECT"
1700 PRINT  "-----------------------------------------------"
1710 PRINT
1720 J9=J9+1
1730 RETURN
1740 END
```

Figure 3. Dead time compensation computer listing.

Table 1. Sample output of DTC.

DATA SUMMARY

PROCESS TRANSFER FUNCTION: 3 / (3 * S + 1)
P1 CONTROLLER: 10 * (1 + 1 / 0.2 * S)
DEAD TIME: TAU = 1

SET-POINT STEP CHANGE : 3
LOAD STEP CHANGE : 0

INTEGRATION STEP : 0.01
FINISHING TIME : 2
PRINTING INTERVAL : 0.1

TIME	C PLANT OUTPUT	C1 COMPENSATED OUTPUT	C1-C COMPENSATOR EFFECT
0.1	0	2.35612	2.35612
0.2	0	3.34592	3.34592
0.3	0	3.57199	3.57199
0.4	0	3.47188	3.47188
0.5	0	3.29299	3.29299
0.6	0	3.14271	3.14271
0.7	0	3.04829	3.04829
0.8	0	3.0017	3.0017
0.9	0	2.98553	2.98553
1.0	0	2.98455	2.98455
1.1	2.35612	2.98897	0.632854
1.2	3.34592	2.99386	−0.352061
1.3	3.57199	2.99739	−2.574602
1.4	3.47188	2.99938	−0.4725
1.5	3.29299	3.00024	−0.292756
1.6	3.14271	3.00045	−0.142254
1.7	3.04829	3.00039	−0.0478992
1.8	3.0017	3.00024	−0.00145483
1.9	2.98553	3.00012	0.0145946
2.0	2.98455	3.00004	0.0154953

SUGGESTED COMPLEMENTARY READING

1. Smith, O. J. M., *ISA Journal*, *6*, 28–33 (Feb 1959).
2. Lipfer, D. E. & M. W. Oglesby, ISA Journal, *8*, 53–57 (Nov. 1961).
3. Perry, R. H. & C. H. Chilton, "Chemical Engineers Handbook," Chapter 22, 5th Ed., McGraw Hill (1973).

STUDY PROBLEMS

1. List which of the causes for dead time described in the introduction may contribute to θ_D and which to θ_p.

2. What effect may have each of θ_D and θ_p on the stability of the control system in Figure 1?

3. Is a feedback control system that contains dead time necessarily unstable?

4. What effect, if any, will the compensation have on the response of the process to changes in set point and to changes in load?

HOMEWORK PROBLEM

1. A product is made by mixing two powders, *A* and *B*. The properties of *B* determine how much *A* should be added to achieve a desired product quality. *B* is not totally uniform, therefore the product quality is controlled by adjusting the amount of *A* added. *A* and *B* are each fed to the mixing tank on a conveyor travelling for approximately 0.5 minutes before reaching the tank. The average residence time in the mixing tank is approximately two minutes. The mass flow rate of *B* is roughly 1000 times that of *A*. It takes approximately three minutes to obtain the results of an analysis of product quality.

   There have been lately complaints about substantial amounts of product being discarded as off-specification. In particular it is noted that every time a new batch of *B* reaches the process, product quality oscillates wildly for some time. There seems to be no satisfactory tuning adjustment of the controller parameters that would avoid that situation and still yield reasonably fast control action. Can this problem be solved by dead time compensation? Formulate the compensation element.

Module A4.7

Interaction Matrix Analysis

T. J. Ward

Clarkson College
Potsdam, New York 13676

OBJECTIVES

Upon completion of this unit, the student should be able to:

1. Appreciate the problem of selecting controlled-manipulative pairs in the common multiloop feedback configuration.
2. Appreciate the interaction complication.
3. Understand and use the relative gain matrix and interaction index as design aids.

PREREQUISITE MATHEMATICAL SKILLS

1. Matrix inversion.
2. Simultaneous solution of algebraic equations.

PREREQUISITE ENGINEERING AND SCIENCE SKILLS

1. Single-loop feedback control concept.

Interaction matrix analysis offers a practical tool to aid in the pairing of manipulative inputs and controlled outputs for the multiloop control configuration. It also offers a measure of the interaction that exists between the loops. As a result, this simple approach can be useful in defining the control problem and predicting interaction difficulties.

THE PROBLEM

Let us consider a process with several manipulative inputs (the m variables) and several variables to be controlled (the c variables), as shown in Figure 1. If three feedback loops are to be designed for this process, one question should be asked, "Which manipulative input should be aligned (or paired) with which controlled variable in each loop?" This is not always easy to answer because two possibly-conflicting goals are desirable:

a) The output c_1, for example, should be controlled by manipulating the one input capable of exerting the greatest influence on c_1.

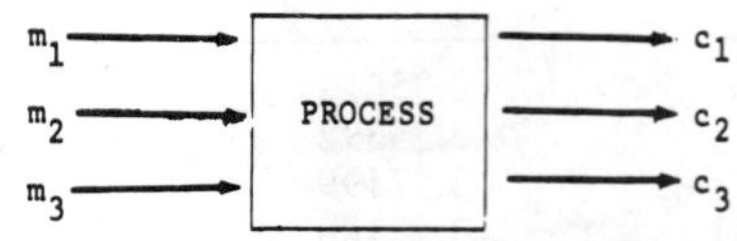

Figure 1. Multivariable process.

b) All of the other manipulative inputs should have no effect on c_1, in order to minimize interactions between the loops.

If there are n loops to be designed, there are $n!$ ways to align the variables in input-output pairs. One simple approach might be to compare the the open-loop gains relating each m input to each of the outputs to be controlled. However, different dimensions are involved and the gains are more meaningful if they are scaled to some normalized dimensionless form. The relative gain matrix provides one approach to this problem.

RELATIVE GAIN MATRIX

There are two types of open-loop gains that one can obtain to relate a pair of variables, m_j and c_i. The gain $\partial c_i/\partial m_j$ can be evaluated with either:

a) all other m held constant
b) all other c held constant

The first case, designated $(\partial c_i/\partial m_j)_m$, corresponds to the usual open-loop situation where none of the other manipulative inputs are changed. The second case, $(\partial c_i/\partial m_j)_c$, would be that in which all of the other controlled outputs are held constant by control action. In other words, all of the other loops would be closed.

The ratio of these two gain quantities is called the *relative loop gain* λ_{ij}, given by

$$\lambda_{ij} = \frac{(\partial c_i/\partial m_j)_m}{(\partial c_i/\partial m_j)_c} \tag{1}$$

These relative loop gains can be arranged conveniently in a square array with the typical element λ_{ij} in the ith row and jth column (Figure 2). This array is called the *relative gain matrix*. The sum of λ elements in any row or column of the array is unity. These two properties can be written as

$$\sum_{i=1}^{n} \lambda_{ij} = 1 \qquad (j = \text{constant}) \tag{2}$$

	m_1	m_2	...	m_j	...	m_n
c_1	λ_{11}	λ_{12}	....			λ_{1n}
c_2	λ_{21}	λ_{22}				.
.						.
c_i				λ_{ij}		.
.						.
c_n	λ_{n1}	....				λ_{nn}

Figure 2.

$$\sum_{j=1}^{n} \lambda_{ij} = 1 \qquad (i = \text{constant}) \tag{3}$$

Before the relative gain matrix can be used, one has to devise procedures for evaluating the two types of gain terms involved in each element. Surprisingly, one really has to find only *one* set of gain terms—either the $(\partial c_i/\partial m_m)_m$ or the $(\partial c_i/\partial m_j)_c$. The reason is that matrix algebra can be used to relate the two quantities.

RELATIONSHIP BETWEEN THE GAINS

a) Known: $(\partial c_i/\partial m_j)_m$

Let us consider the case where all of the $(\partial c_i/\partial m_j)_m$ have been determined. Arrange these known gains as a square matrix **A**, with the typical element defined as $a_{ij} = (\partial c_i/\partial m_j)_m$.

$$\mathbf{A} = \begin{bmatrix} (\partial c_1/\partial m_1)_m \; (\partial c_1/\partial m_2) \ldots (\partial c_1/\partial m_n)_m \\ (\partial c_2/\partial m_1)_m \\ \cdot \\ \cdot \\ \cdot \\ (\partial c_n/\partial m_1)_m \qquad\qquad (\partial c_n/\partial m_n)_m \end{bmatrix} \tag{4}$$

A matrix **B**, composed of the elements $b_{ij} = (\partial m_j/\partial c_i)_c$ is given by the transpose of the inverse of **A**

$$\mathbf{B} = [\mathbf{A}^{-1}]^T \tag{5}$$

In other words, these matrix operations give

$$\mathbf{B} = \begin{bmatrix} (\partial m_1/\partial c_1)_c \; (\partial m_2/\partial c_1)_c \ldots (\partial m_n/\partial c_1)_c \\ (\partial m_1/\partial c_2)_c \\ \cdot \\ \cdot \\ \cdot \\ (\partial m_1/\partial c_n)_c \qquad \ldots (\partial m_n/\partial c_n)_c \end{bmatrix} \tag{6}$$

Now the elements λ_{ij} in the relative gain matrix can be found by multiplying *corresponding elements* in the A and **B** arrays:

$$\lambda_{ij} = a_{ij} b_{ij} \tag{7}$$

For example,

$$\lambda_{21} = a_{21} b_{21} \tag{8}$$

$$= (\partial c_2/\partial m_1)_m \cdot (\partial m_1/\partial c_2)_c \tag{9}$$

$$= \frac{(\partial c_2/\partial m_1)_m}{(\partial c_2/\partial m_1)_c} \tag{10}$$

b) Known: $(\partial c_i/\partial m_j)_m$

Now let us look at the case where the $(\partial c_i/\partial m_j)_c$ have been determined. Take the reciprocal of each element to give $(\partial m_j/\partial c_i)_c$. These are arranged to give the **B** matrix directly. The transpose of the inverse **B** is the **A** matrix

$$\mathbf{A} = [\mathbf{B}^{-1}]^T \tag{11}$$

As above, the product of corresponding elements a_{ij} and b_{ij} gives the λ_{ij} element of the relative gain matrix.

Now the task has been simplified. Since one can use a matrix calculation to relate the two types of gains, only one type of gain needs to be found. Just how is the gain evaluated? If one has an existing process, experimental tests can be run. Better yet, if the steady-state equations relating the m and c variables are known, one can make an analytical (or numerical) evaluation of the gains. Let us look at each approach in the following section.

EXPERIMENTAL GAIN EVALUATION

For an existing process, one might want to determine the gains in order to assess troublesome interaction effects. There are two types of experimental gain determination that could be made. Let us consider each.

a) *Open-loop.* If all of the loops are placed in a manual or open-loop mode at the design steady-state, then we can make a small change Δm_j in one input and measure the corresponding steady-state changes Δc_1, Δc_2, ... Δc_n in all of the controlled outputs. Since $\Delta c/\Delta m \simeq \partial c/\partial m$, then the test gives $(\partial c_1/\partial m_j)_m$, $(\partial c_2/\partial m_j)_m$, ... After this test, the variable m_j is returned to its steady-state value. The same test procedure is repeated for each of the m variables in turn to give the elements of the A matrix.

b) *Closed-loop.* If all of the loops are closed and the process is at the design steady-state, the first step would be to place one loop on manual mode. The corresponding input m_j can be incremented slightly and the steady-state changes in all of the c variables noted to give $(\partial c_1/\partial m_j)_c$, $(\partial c_2/\partial m_j)_c$, ... Then m_j is returned to its steady-state and the loop returned to automatic. This is repeated for each loop in turn to give the reciprocals of all the elements of the **B** matrix.

An alternate closed-loop procedure would be to keep all loops closed and make a small setpoint change in one of the controlled variables c_i. The steady-state change Δc_i and the corresponding steady-state changes Δm_1, Δm_2, ... in all of the manipulative inputs would be noted. Then c_i is returned to its original steady-state. This can be repeated for each of the c variables to give the elements of the **B** matrix directly.

NUMERICAL GAIN EVALUATION

If the algebraic equations relating steady-state inputs and outputs are known, we can estimate the gains without any experiments. All we do is carry out the appropriate partial differentiation of the process equations. However, any processes of practical interest are usually so complex that numerical evaluations of the gains are more useful. Let us outline now what is involved in the numerical evaluation of $(\partial c_i/\partial m_j)_m$ or $(\partial m_j/\partial c_i)_c$:

a) $(\partial c_i/\partial m_j)_m$—The algebraic steady-state process equations are arranged so that each control variable c_i is given as a function of the manipulative inputs and the other controlled variables:

$$c_i = f_i(m_j, c_k) \tag{12}$$

where $1 \leq j \leq n$ and $1 \leq k \leq n$, with the restriction that $k \neq i$. The steady-state values of all c and m variables are determined by solving the system equations and utilizing the specified steady-state data.

Then one of the m inputs (say m_j) is changed slightly. Holding all other m constant, we can solve the system of equations for the new set of c values. Then m_j is returned to its original steady-state values. This procedure is repeated for each of the m inputs. The $\Delta c/\Delta m$ values obtained are estimates of the elements of the **A** matrix.

b) $(\partial m_j/\partial c_i)_c$—The algebraic steady-state process equations are arranged so that each manipulative input m_j is given as a function of the controlled variables and the other manipulative inputs:

$$m_j = f_j(c_i, m_k) \tag{13}$$

where $1 \leq i \leq n$ and $1 \leq k \leq n$, with the restriction that $k \neq j$. Using the specified steady-state values, solve the equations for the steady-state values of all the m and c variables.

Then, one of the c variables (say c_i) is incremented slightly. Holding all other c variables constant, one solves the equations for the new set of m values. Then c_i is returned to its original steady-state. This procedure is repeated for each of the c variables. The $\Delta m/\Delta c$ values that can be calculated are the elements of the **B** matrix.

Since the specified steady-state is usually given as values of the c variables rather than values of the m variables, it is often more convenient to find **B** matrix first and then use the matrix relationship to find the **A** matrix.

EXAMPLE PROBLEM

Let us consider an example problem based on Figure 3 and an illustration of Shinskey (*2*). The object here is to blend side-chained hydrocarbons (S) with linear hydrocarbons (L) to produce a product P.

Figure 3. Mixing example.

Assume that the flow rates are also given by the symbols S, L, and P, so that $P = S + L$. The steady-state concentrations of side-chain components in the product is $x = S/(S+L)$. It is desired to control both the product rate P and concentration x by manipulating the input flows S and L. How should one align the variables? In other words, which configuration in Figure 4 is more desirable?

For this example, let us assume first that the specified steady-state is given by $P = 4$ and $x = 0.25$. The steady state process model is given by the pair of equations

$$P = S + L \tag{14}$$

$$x = \frac{S}{S+L} \tag{15}$$

For calculation purposes, order the controlled and manipulative variables as $c_1 = P$, $c_2 = x$, $m_1 = S$, and $m_2 = L$.

a) *Determination of* $(\partial c_i/\partial m_j)_m$—The process equations are already in the form of Equation 12. Solve for the steady-state values of the remaining unknowns to give $S = 1$ and $L = 3$.

First, increment S by selecting $\Delta S = 0.01$ to give $S = 1.01$. Holding $L = 3$, solve for the new steady-state as $P = 4.01$ and $x = 0.25187$. Now one can identify the following gains:

$$(\partial c_1/\partial m_1)_{m2} \simeq \frac{\Delta P}{\Delta S} = \frac{0.01}{0.01} = 1 \tag{16}$$

$$(\partial c_2/\partial m_1)_{m2} \simeq \frac{\Delta x}{\Delta S} = \frac{0.00187}{0.01} = 0.187 \tag{17}$$

Next, S is returned to its original steady-state and L is incremented so that $\Delta L = 0.01$ and $L = 3.01$. Holding $S = 1$, solve for the new steady-state as $P = 4.01$ and $x = 0.249377$. Now, calculate two more gains as:

$$(\partial c_1/\partial m_2)_{m1} \simeq \frac{\Delta P}{\Delta L} = \frac{0.01}{0.01} = 1 \tag{18}$$

$$(\partial c_2/\partial m_2)_{m1} \simeq \frac{\Delta x}{\Delta L} = \frac{-0.000623}{0.01} = -0.0623 \tag{19}$$

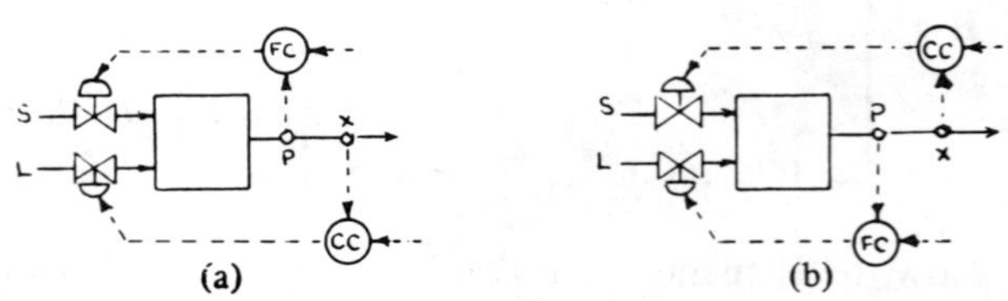

Figure 4. Two control configurations.

This gives the **A** matrix as

$$\mathbf{A}=\begin{pmatrix}1 & 1\\ 0.187 & -0.0623\end{pmatrix} \tag{20}$$

Since $\mathbf{B}=[\mathbf{A}^{-1}]^T$, solve for B to obtain

$$\mathbf{B}=\begin{pmatrix}0.25 & 0.75\\ 4 & -4\end{pmatrix} \tag{21}$$

Now $\lambda_{ij}=(a_{ij})(b_{ij})$, so that the relative gain matrix is

	m_1	m_2
c_1	0.25	0.75
c_2	0.75	0.25

b) *Determination of* $(\partial m_j/\partial c_i)_c$—First, rearrange the process equations to the form of Equation 13 to give:

$$\begin{aligned} S&=Px\\ L&=P(1-x)\end{aligned} \tag{22}$$

Increment P by 0.01 to give $P=4.01$. Holding x constant, solve for the new steady state to obtain $S=1.0025$ and $L=3.0075$. From this, calculate

$$(\partial m_1/\partial c_1)_c\simeq\frac{\Delta S}{\Delta P}=\frac{0.0025}{0.01}=0.25$$

$$(\partial m_2/\partial c_1)_c\simeq\frac{\Delta L}{\Delta P}=\frac{0.0075}{0.01}=0.75$$

Next, P is returned to its original value and x is incremented by 0.001 to give $x=0.251$. Holding P constant, solve the equations to obtain $S=1.004$ and $L=2.996$. From this, one gets

$$(\partial m_1/\partial c_2)_c\simeq\frac{\Delta S}{\Delta x}=\frac{0.004}{0.001}=4$$

$$(\partial m_2/\partial c_2)_c\simeq\frac{\Delta L}{\Delta x}=\frac{-0.004}{0.001}=-4$$

Now we have obtained the elements of the **B** matrix as

$$\mathbf{B}=\begin{pmatrix}0.25 & 0.75\\ 4 & -4\end{pmatrix}$$

Since $\mathbf{A}=[B^{-1}]^T$, then

$$\mathbf{A}=\begin{pmatrix}1 & 1\\ 0.187 & -0.0623\end{pmatrix}$$

Note that these are the same matrices obtained by procedure (a) and give the same relative gain matrix.

While the two procedures for calculating the relative gain matrix have been illustrated, the question of how to pair the variables still hasn't been answered. Let us see how the relative gain matrix can provide the answer.

USE OF THE RELATIVE GAIN MATRIX

The relative gain elements are measures of the influence of the manipulative inputs on the controlled variables, with the larger (more positive) values indicating the greater effect. This suggests that c_1, for example, should be controlled by manipulating the m variable giving the most positive gain in the first row of the relative gain matrix. Returning to our example problem where:

		S	L
		m_1	m_2
P	c_1	0.25	0.75
x	c_2	0.75	0.25

one would select L as the variable to pair with P and S as the variable to pair with x.* Once these pairings are made, the control problem is specified. One can also estimate the induced interaction effects by examining the gain elements relating the m of one loop to the c of another loop. For the above example, it can readily be seen that the interaction-induced effects are about 1/3 of the control effort effects. It is possible to generalize this notion in terms of an interaction index matrix **I**. This approach of Nisenfeld and Schultz (*4*) is discussed briefly in the next section.

INTERACTION INDEX MATRIX

The interaction index is a relative measure of the interaction gains. Since the sum of the relative gain elements in any row or column is unity, one can define the interaction gains associated with a gain element λ_{ij} as $(1-\lambda_{ij})$. Then an interaction index can be defined:

$$\mathring{1}_{ij}=\frac{1-\lambda_{ij}}{\lambda_{ij}}$$

and the matrix composed of such elements as the interaction index matrix **I**.

For our example problem, the **I** matrix would be

$$\mathbf{I}=\begin{pmatrix}3 & 0.333\\ 0.333 & 3\end{pmatrix}$$

Any variable pair for which the interaction index is greater than one is an unacceptable loop pairing and would be rejected. For example, in our problem the $S-M$ and $L-x$ pairing would be unacceptable, as already concluded in the previous section.

CONCLUSION

Computer programs could be developed to handle the computations if one was going to do this kind of calculation often for complex systems. Some industrial organizations use such programs.

One should realize that this approach involves steady-state equations and neglects the dynamic response effects. However, it is simple, practical, and useful as an aid in specifying the control problem in many cases.

*In the Homework Problem 2, you may find that a different pairing would be selected when x = 0.75 instead of 0.25.

LITERATURE CITED

1. Bristol, E. H., "On a New Measure of Interactions for Multivariable Process Control," *IEEE Trans. Auto. Control, AC-11*, 1, 133 (1966).
2. Shinskey, F. G., "Process Control Systems," pp. 188–192, McGraw-Hill, New York (1967).
3. Nisenfeld, A. E., and Stravinski, C. A., "Feedforward Control For Azeotropic Distillations," *Chem. Eng.*, *75*, 20, 227–236 (1968).
4. Nisenfeld, A. E., and H. M. Schultz, "Interaction Analysis Applied to Control System Design," *Instrumentation Technology*, 52–57 (April 1971).
5. Smith, C. L., "Digital Computer Process Control," pp. 212–217, Intext, Scranton (1972).

STUDY PROBLEMS

1. What two considerations are important in deciding which manipulation is going to be used to control a given process variable? Can you think of others as well?
2. Why are interactions between the loops a problem?
3. Manipulative inputs are generally constrained. For example, a valve can never be more fully-open or less than fully-closed. How would constraints affect the gain matrix analysis?
4. What is the transpose of a matrix?
5. Give a simple defining equation for the inverse of a matrix.
6. What is the difference between $(\partial c_i/\partial m_j)_c$ and $(\partial c_i/\partial m_j)_m$?

HOMEWORK PROBLEMS

1. Shinskey (*2*) analyzes the same mixer example used in the notes. For this simple case, the necessary partial differentiation can be done analytically. Derive the gain matrix elements as functions of the composition x. Compare with the numerical calculations for $x=0.25$.
2. In the example, the steady-state concentration was 0.25. Do the numerical calculations for a) $x=0.75$ and b) $x=0.25$. What do you conclude from your results? Compare with the results that could be calculated in Homework Problem 1.

Module A4.8

Selection of Control Measurements

Marshall E. Findley

University of Missouri-Rolla
Rolla, Missouri 65401

OBJECTIVES

At the completion of this module the student should be able to:

1. Determine the limitations on the number of control variables in a simple or complex process by analysis of the degrees of freedom.
2. Select a set of variables for measurement and control based on desired process performance characteristics without violating limitations on a process unit basis or an overall process basis.
3. Select variables to control for steady-state or near steady-state maintenance, for insuring certain assumptions or desired design or operating conditions, and for prevention of unnecessary upsets.
4. Select manipulated variables for control loops, and modify control variable selection to provide stable control loops.

PREREQUISITE MATHEMATICAL SKILLS

1. An understanding of simultaneous equations.

PREREQUISITE ENGINEERING AND SCIENCE SKILLS

1. Material and energy balances.
2. Rate relationships.
3. Signal flow diagrams, control loop functions and control nomenclature.

One of the most common problems faced by chemical engineers in the field of control is in the design of control systems for new or modified processes. In other words, the engineer may have to answer the question, Given the flow sheet and process information, where do we place the measuring, controlling, and manipulating instruments or equipment?

While control schemes may be analyzed unit of equipment by unit, such unit analyses will depend on which variables are controlled in other units. This and the fact that overall process performance is the major reason for control indicate that there are advantages in studying the control of all units simultaneously.

An important factor in overall process control is the coordination of the various control systems and the physical process relationships, so that two (or more) controllers are not trying to control two variables at set points which are physically imcompatible or mutually exclusive. If this happens we will have two controllers fighting each other to attain their set points, and one will probably dominate fixing the other variable at an unplanned value. This module provides a method of selecting variables for measurement and control based on the degrees of freedom of individual units and the complete multi-unit process.

CRITERIA FOR SELECTION OF VARIABLES FOR MEASUREMENT AND CONTROL

Before selecting control variables for a process, the major purpose and requirements of the process should be determined, because the purpose of the control system is to maintain conditions that achieve the process purpose and satisfy process requirements. Quality of product, rates of production, and minimum costs of production are the types of factors often involved in process purposes and requirements, usually with one of these factors more critical than the others.

In most processes it is important to maintain constant conditions or a certain desired pattern of change occurring over a relatively long time. The primary purpose of the control system is to monitor the important conditions, and when outside uncontrolled variations change these conditions, the control system should rapidly restore the desired conditions by manipulating an appropriate process variable. Thus the dynamics of the process and control system are critical in maintaining a steady or slowly changing desired process condition. When a variable is selected to measure and control it is also necessary to select a variable to manipulate so that the controller-manipulated variable-process loop will respond with satisfactory dynamics. Although the consideration of process dynamics for each control loop is necessary, this study will concentrate on selecting the desired control variables and the limitations on these selections, which depends primarily on steady-state or near steady-state conditions. In addition, each control system selected should be satisfactorily operable, and all the control systems should be operable simultaneously under the desired process conditions. To achieve satisfactory overall process performance, the overall process control should provide the following desired characteristics:

1. Provision should be made to control variables with a major effect on the process at desired values. In addition, steady-state or other necessary or desired conditions should be provided by controlling appropriate variables. Such variables are likely to include either feed or product rates, variables which assure performance (concentrations or conditions), ratios or flows of recycle and intermediate streams, and hold-up variables.

2. Each controlled variable should be connected in a control loop with a manipulated variable that will have a definite, reliable, and consistent effect on the controlled variable. Usually the simpler the loop the better.

3. The dynamics of each control loop under operating conditions should be such that process performance is not disturbed.

4. There should be a minimum of interaction between loops, unless specially desired, and when interaction exists, alternate possibilities should be considered to keep interactions confined to effects of controlled variables or to variables with small and slow fluctuations as much as possible. If possible, avoid control loops which depend on another control loop to provide the desired response.

5. In processes with inherent fluctuations, arrange controls or the process to isolate, damp out, or suppress the effects of these fluctuations on the process performance.

However, even though we know which variables we would like to control, there are limitations on the number and combinations of variables that can be controlled.

DETERMINING THE DEGREES OF FREEDOM FOR CONTROL VARIABLE SELECTION

In order to determine how many and which variables may be controlled, one needs to know the number of degrees of freedom in the process system.

$$N_F = N_V - N_E$$

where N_F equals the degrees of freedom, which equals the maximum number of independent control variables and variables fixed outside the system studied.

N_V = number of variables

N_E = number of equations physically relating the N_V variables

The system of equations and variables can be as simple or as complex as desired, providing all independent equations relating the selected variables in the desired operation are included. Desired design condition equations should not be included unless they are physically determined. Such design conditions should be considered when selecting control variables.

Where steady state is desired in process operations, there are several ways to proceed. One way is to provide material balance controls for steady state (levels for liquids, pressures for gases) if necessary, after determining N_F for unsteady-state total flow equations. Then the process degrees of freedom can be determined from steady-state equations (if pressure is controlled for steady state, consider it fixed outside for the process). A second approach is to consider all equations as unsteady-state with the necessary hold-up variables (levels and pressures). A third method is to assume steady state and possible other assumptions, such as separation or concentrations, in the main set of process equations, determine the process N_F, and then add steady-state controls, and any controls necessary for other assumptions. The number of control variables allowed by the degrees of freedom in a unit or multi-unit system is the number of degrees of freedom minus the number of variables fixed outside the system of variables under study, often by a previous process or by the surroundings, and sometimes by a following unit (for example, when an outflow is fixed by an inflow manipulation in a following process). Note that "fixed," does not imply a constant variable but one determined by controls, surroundings, or process performance outside the system under study. Almost always, compositions of inflows will be fixed by the source of the stream, and in many cases inflow rates will be fixed by the previous process. For example, consider the simple evaporator in Figure 1. Let F and X_F be fixed by a previous unit. The desired design conditions are $P = 1.5$ atm and $X_B = 2_F$ (or $B = 0.5\ F$).

If steady-state controls are selected first, the liquid hold-up could be fixed by a level controller manipulating outflow B (since F is fixed it cannot be manipulated). One could also fix the vapor hold-up by controlling pressure with manipulation of a vapor outflow valve. The next step is the control of process performance, and this can be studied with steady-state flow equations.

One could write the following equations:

Total Material Balance

$$F = B + D \tag{1}$$

Component Balance (2 *components*)

$$FX_F \doteq BX_B + DX_D \tag{2}$$

$$F(1 - X_F) = B(1 - X_B) + D(1 - X_B) \tag{3}$$

where X = mol fraction solute, and $(1 - X)$ = mol fraction solvent. Note that Equation 3 can be derived from Equations 1 and 2 so that it is not independent and should not be counted in N_E.

Energy Balance

$$FH_F + W_S \lambda_S = (F - B)\lambda_T + BH_B \tag{4}$$

where H_F and λ_T and H_B are functions of T. It is not

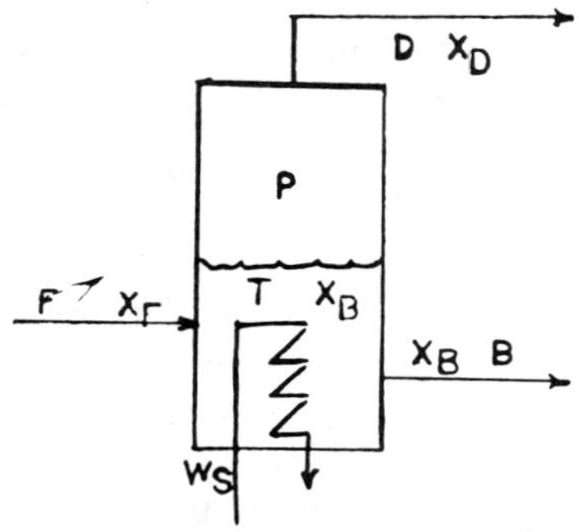

Figure 1. Evaporator with important variables.

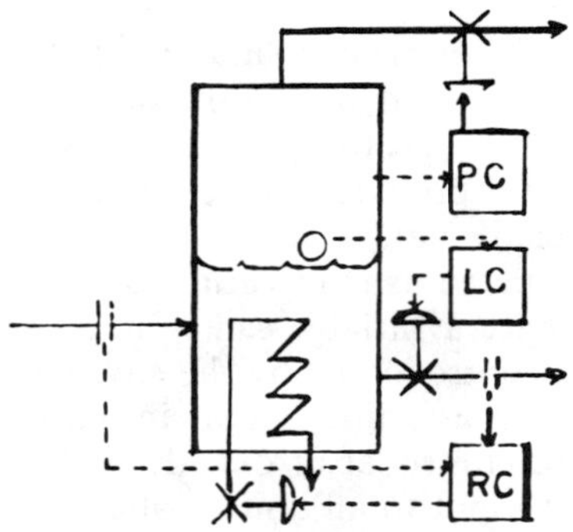

Figure 2. Evaporator illustrating controls; PC is the pressure controller, LC is the level controller, and RC is the ratio controller.

necessary to know the exact equation; one can simply write $B=\psi_1(W_S, T, F)$.

Equilibrium

$$T=\psi_2(P, X_B) \quad (5)$$

The above function represents the vapor pressure versus temperature relationship at various mol fractions of solute.

$$X_D=0 \quad (6)$$

This unit could include heat transfer equations with steam temperature and pressure, and other equations, but it is not necessary if the exchanger is adequate and one does not wish to study control of other variables.

By inspection of only the total material balance, Equation 1, there are 3 variables, 1 equation, and F is fixed, so a maximum of one of B and D can be controlled.

By inspection of the total and component balances and $X_D=0$ (Equations 1, 2, and 6), one can see that among F, D, B, X_F, X_B, and X_D, only one of these variables can be controlled (6 variables-3 equations-2 fixed variables). This provides almost all of the information needed for the problem, but alternatively, the whole system could be considered.

There are five independent equations and 9 variables—F, D, B, X_F, X_D, X_B, W_S, T, P—so $N_F=9-5=$degrees of freedom. However, F, X_F are fixed by a previous unit and must be independent of these equations. Also, P has been controlled to maintain vapor hold-up so it is fixed outside this system of equations. Thus there is only one independent variable left to control, with five physical relationships that will fix the remaining five variables.

The choice is now to choose the one control variable. Based on the desired design condition of $B=1/2\,F$, B can be selected to control at the desired ratio with F. If desired, $R=B/F$ could be added to the equations with R an additional variable and this would not change the degrees of freedom or the previous analysis. The control variable could then be R. Since the outflows are already manipulated to maintain steady state, W_S would be the obvious variable to manipulate to maintain the measured and controlled ratio of B/F. This control system is shown in Figure 2. Note that flow or ratio control of B, manipulated by level control, in effect transfers level manipulation to W_S and provides flow manipulation through the level control valve. Thus the level control system must work for flow control to be effective, and this is probably less desirable than direct flow control of B and level control manipulation of W_S.

Another procedure would be to assume steady state, and consider the same five independent equations and nine variables, to get $N_F=4$. Since F and X_F are fixed elements elsewhere, P and B (as a ratio) could be selected as the 2 control variables, possibly by manipulating D and B with valves. Then steady-state control could be added by a level controller manipulating steam flow W_S, with P already controlled. Of course, the same manipulated variables as in the first procedure could also be used.

These methods can be considered as a signal flow block with independent variables as inputs and the degrees of freedom noted. Control variables must be inputs and manipulated variables should be outputs unless also measured and controlled by manipulating some other variable. The evaporator could be shown as in Figure 3a, before control variable selection, and as in 3b, after control variable selection; with F, C, and

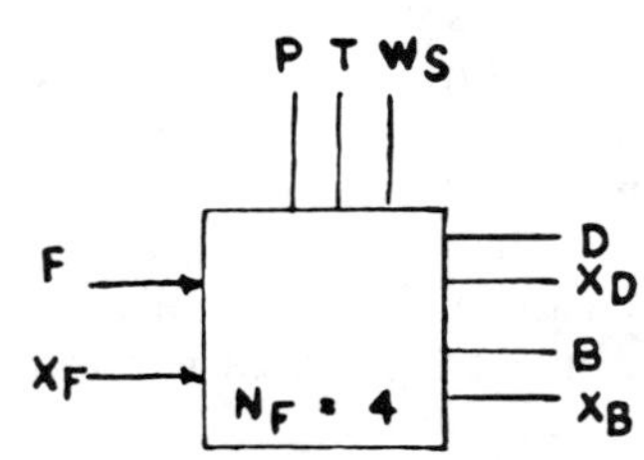

Figure 3a. Block diagram of evaporator before control variable selection.

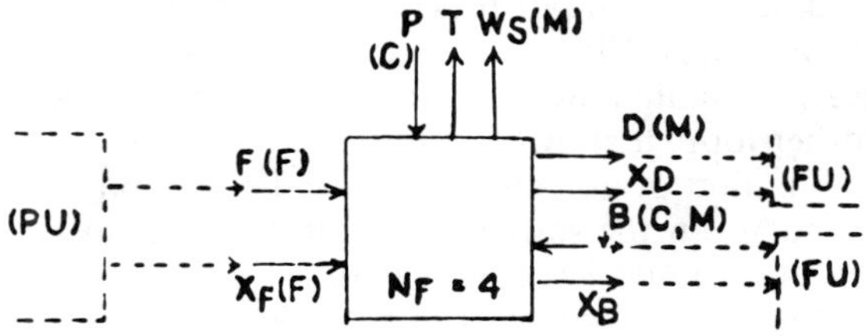

Figure 3b. Evaporator, after control variable selection, showing variables, degrees of freedom, and independent variables.

M denoting fixed, control, and manipulated variables, and (PU) and (FU) denoting the previous and following unit, respectively. Steady-state level control is not indicated if provided for separately, and if N_F is based on steady state. The variables fixed elsewhere can be considered as load variables in dynamic studies if desired. Since a variable can be fixed by only one unit or one controller, or by the environment, any variable shown as a line can be an output from only one block or controller or outside, and must be an independent input to all units it affects. This block type of representation gives a convenient way of connecting units into an overall process. Thus, a system of blocks can represent units or systems of equations, with lines as variables. For example, Figure 3b could be combined with a previous unit block determining F and X_F as outputs and with following unit blocks where D, X_D, X_B, and B are fixed inputs from the evaporator. The control variable B should form inputs to both the evaporator and a following unit, since it is determined by a controller.

SELECTION OF CONTROL VARIABLES BY BLOCKS

If a system consists of a number of processes and units, each of which is simpler than the whole, there is an advantage in considering the system as a number of blocks and arranging the variables, equations, and degrees of freedom according to units. It is better to consider process unit equations rather than overall equations, because the variables are to be studied and degrees of freedom determined initially based on units or blocks. The design conditions on the overall process will be helpful in selecting control variables.

In each process unit, the number of variables and independent equations are determined in that unit, and the degrees of freedom is determined by $N_F = N_V - N_E$. An equation should be used in only one block. A block diagram can be made including each unit as a block with its degrees of freedom noted. Each variable is represented by a line, branched as necessary, to connect with any block which that variable affects (or any block in whose equations that variable appears). This is similar to the procedure used in design strategy by Rudd and Watson (7). Originally, directions should not be shown, but as controlled variables are assigned, the directions are indicated.

A variable that is fixed outside the overall process or controlled, will be an independent input to all units in which it is involved, and a variable that is not controlled or fixed will be a dependent output of only one unit and an input to all other units in which it is involved, normally in the direction of the process streams. Manipulated variables not controlled by another loop, if shown, will be outputs of the unit or combination of units where manipulated, and are dependent on the state of the unit. In the final system each unit should have only as many input arrows as it has degrees of freedom. The total degrees of freedom can be obtained as the sum of the unit degrees of freedom minus the number of times variables are connected to more than one block. Briefly the procedure is as follows:

1. Separate the overall process into blocks and determine the degrees of freedom in each.
2. Draw one line for each variable, connected to all blocks in whose equations that variable appears, without directions.
3. Make fixed variables inputs to all units, and temporarily assume composition and flow variables are each an output of an upstream unit and inputs to all others.
4. Select important control variables and manipulations, with control variables inputs to all units. If necessary, readjust directions of stream variables so that no unit will have more inputs than degrees of freedom; when inputs equal N_F make all other variables from a block outputs and inputs to all others. Select less important control variables in each unit and any variables to be allowed to depend on the environment as inputs until all blocks have inputs equal to N_F.
5. If steady-state was assumed in number one without hold-up variable selection, steady-state control variables and manipulations should now be added.

Distillation Column

As an example, a distillation column can be broken down into individual trays, as shown in Figures 4a-c.

The degrees of freedom can be arrived at by considering a single tray, as in Figure 4 and the corresponding work in Table 1. To summarize: In each tray there are C material balances on component flows, and C equilibrium relationships where C is the

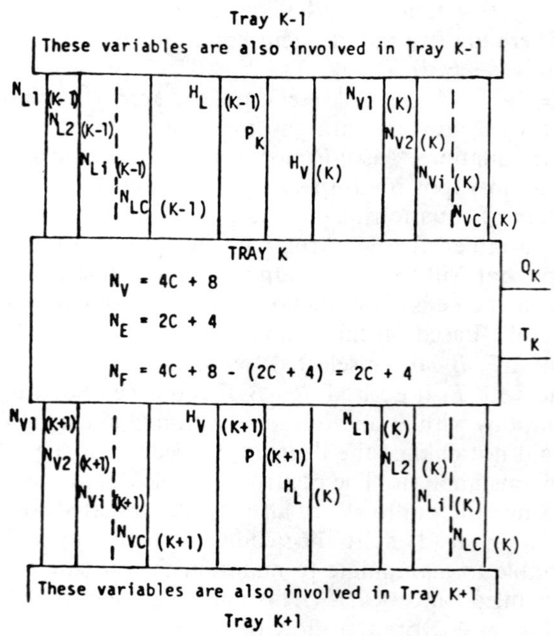

Figure 4. Individual distillation tray block diagram

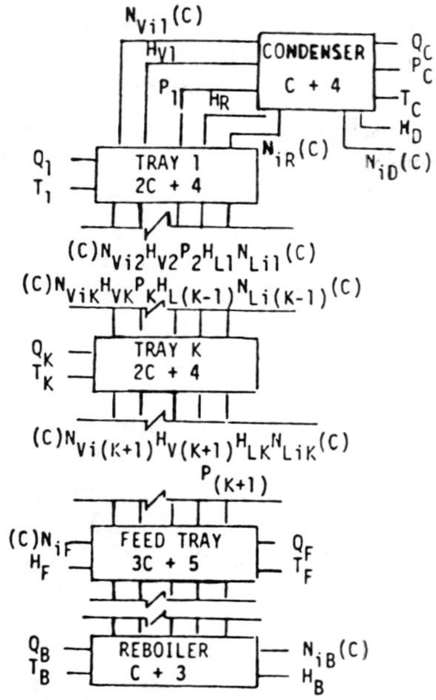

Figure 4a.

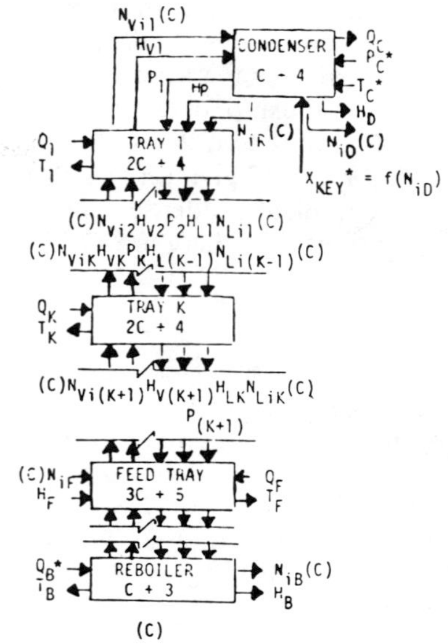

*Controlled variable

Figure 4c.

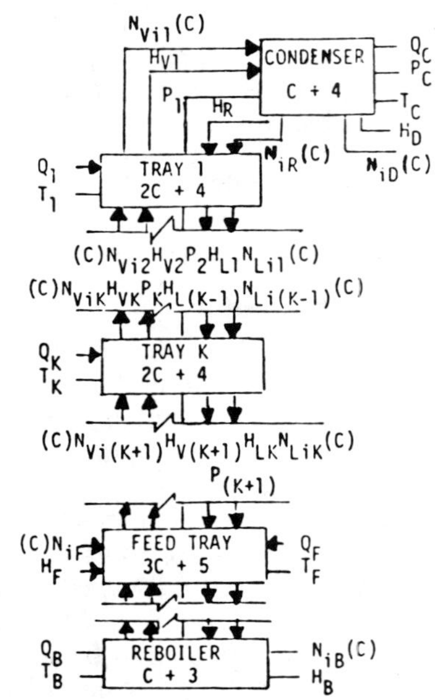

Figure 4b.

Table 1.

Equations Each Tray

C Material Balances (C = No. of components):

$$N_{Li(K-1)} + N_{Vi(K+1)} = N_{Vi(K)} + N_{Li(K)} \qquad i = 1 \text{ to } C$$

C Equilibrium Equations:

$$\frac{N_{Vi(K)}}{\Sigma N_{Vi(K)}} = K_i \frac{N_{Li(K)}}{\Sigma N_{Li(K)}} \qquad i = 1 \text{ to } C$$

K_i is a function of T_K, P_K, Composition Tray K

2 Enthalpy Equations:

$$H_{L(K)} = \Sigma \bar{H}_{Li(K)} \qquad i = 1 \text{ to } C$$

$$H_{V(K)} = \Sigma \bar{H}_{Vi(K)} \qquad i = 1 \text{ to } C$$

$H_{Li(K)}$ and $H_{Vi(K)}$ are functions of T_K, P_K, Comp. K

1 Enthalpy Equation:

$$H_{L(K)} + H_{V(K)} = H_{V(K+1)} + H_{L(K-1)} + Q_K$$

1 Pressure Drop Equation:

$$P_{(K+1)} - P_{(K)} = C_{\text{TRAY}}(\Sigma N_{Vi(K+1)}) + h_K$$

C_{TRAY} and h_K are constants at steady operation

$2C + 4$ = *Number of Equations*

number of components. Two equations relate vapor enthalpy and liquid enthalpies of leaving streams to compositions, pressure, and temperature. There is one enthalpy balance and one relationship of pressure to vapor flow and the pressure in either the next lower or upper tray (assume the lower tray for this example). This is $2C+4$ equations for each tray. The variables are C for each of the 4 streams (2 in, 2 out), 4 enthalpies, (2 liquid, 2 vapor), pressure P_k, pressure in the lower tray P_{k+1}, heat transferred to the tray Q_k, and temperature T_k, or $4C+8$ variables. Thus there are $C+8-(2C+4)=2C+4$ degrees of freedom for each tray.

The feed tray has the same number of equations but $C+1$ additional variables (feed of each component plus enthalpy), and thus has $3C+5$ degrees of freedom. The reboiler has one less equation, there is no pressure relation with a lower tray, and only 3 streams. Thus the number variables is $3C+6$ ($3C$ component variables, 3 enthalpies, P_B, T_B, Q_B and no P_{B+1}) and the number of equations is $2C+3$. This gives $C+3$ degrees of freedom.

In a partial condenser, there are C material balances and C equilibrium equations, 2 enthalpy equations as functions of flows and temperature, 1 enthalpy balance, and 1 pressure relationship, or $2C+4$ equations. Variables are $3C$ component flows, 3 enthalpies, P_C, T_C, Q_C, and P_1 or $3C+7$ giving $C+3$ degrees of freedom. For a complete condenser, there are no equilbrium equations, but there are $C-1$ equations such as:

$$\frac{N_{iR}}{\Sigma N_{iR}}=\frac{N_{iD}}{\Sigma N_{iD}}$$

(one of this type being derivable from all the rest), or alternatively there are $C-1$ equations such as

$$\left.\frac{N_{iR}}{N_{iD}}\right|_{i=2\text{ to }C}=\frac{N_{1R}}{N_{1D}}$$

(the ratio of reflux to distillate). Otherwise equations and variables are the same, giving $C+4$ degrees of freedom for the complete condensation system, as shown in Figure 4b. In Figure 4b the total degrees of freedom are $(N-2)(2C+4)+3C+5+C+3+C+4-N(C+3)$ connections $=C+N+4$.

The block diagram of a distillation column as shown in Figure 4b for complete condensation can be analyzed for control variable selection by determining first which variables are fixed by factors outside the system or are not considered in the equations. If the feed flows and enthalpy are fixed elsewhere, $C+1$ variables are independent inputs, leaving $N+3$ variables free. If all Q's except Q_C and Q_B are fixed by heat loss relations, then $N-1$ heat values are independent of the system considered, and this leaves 4 variables to control. In terms of the diagram, N_{iF} and H_F are inputs and all Q_K are inputs except Q_C and Q_B. Vapor components and enthalpies can be assumed to flow up and liquid variables down at least tentatively. This gives $2C+2$ inputs for flows and 1 input for Q into each normal tray and allows only 1 other input to the tray (a pressure or temperature). Since each tray involves 2 pressures, the only question left on normal trays is in which direction the pressure variables will be inputs and outputs. The system at this point in the analysis is shown in Figure 4c.

If pressure is fixed by a pressure controller at the condenser, P_C is a controlled input and P_1 will be a condenser output and a tray 1 input, and pressure variables will flow downward. Tray temperatures must then be dependent and outputs.

Under the above assumptions the feed tray is now also fixed, since there are 3 stream inputs, 3 enthalpy inputs and Q and P_{F-1} inputs using up the $3C+5$ degrees of freedom.

In the condenser (assumed complete), there are C component inputs, an enthalpy input, H_{V1}, and P_C as an input, leaving 2 potential control variables. If T_C is left free to vary, Q_C must be fixed, or if it is controlled by manipulating Q_C, it is an independent input and Q_C becomes an output, leaving 1 variable to control. Composition is not a variable considered, but defining

$$X_{\text{key}}=\frac{N_{\text{key }D}}{\Sigma N_{iD}}$$

adds one equation and one variable and doesn't change the degrees of freedom. Thus, X_{key} can be added as a control variable related to product quality, manipulating reflux which is a combination of output variables N_{iR} from the condenser.

In the reboiler, there are $C+1$ liquid inputs from above and a pressure input, leaving 1 variable to control. Q_B or steam flow could be controlled which is almost equivalent to reboil rate, and Q_B becomes the only other input. The complete system noting control variables is shown in Figure 4d.

If it were desired to control the total flow of distillate and distillate composition by manipulating reboil rate and reflux, another variable, $D=\Sigma N_{iD}$, could be a controlled input to the top, and one vapor variable or combination, would have to be an output of the condenser (determined by the requirement of fixed distillate and composition). This would reverse one vapor variable all the way down the column to the reboiler and Q_B, if manipulated, would be reversed to a dependent output at the value required to fix the distillate rate.

If the temperature in the second tray were desired as a control variable, in place of composition, it would become an input to the second tray and another input variable would have to be reversed, probably a flow from the first tray, and this would cause the reversal of a reflux flow variable to an output of the first tray (as it should be if it is manipulated from a column variable). The reflux variable would then be an input to the condenser and the composition variable would be a dependent output instead of a control variable.

Simplified Distillation Block

For convenience, in considering distillation as a block in a large system it is desirable to consider only the major flow variables rather than a detailed analysis. If enthalpy effects are replaced by reflux and reboil rate, the controllable or independently fixed variables are feed variables, 2 compositions and/or flow variables (but not both Bottoms and Distillate flows of any component or total if feed is independent, since $F=B+D$ and $N_F=2$ for F, B, D) and a pressure. If condensation is complete, T_C can be controlled; if condensation is not complete, T_C cannot be controlled if the above variables are all independent. In general, one can ignore T_C and Q_C and consider that in a partial condenser, reflux is a function of Q_C and in a complete condenser T_C or Q_C can be set, with reflux manipulated by a valve.

A distillation block could then be represented by the following equations if pressure changes are ignored.

$$D=\psi_1(R, R_B, P, F, X_F)$$

$$X_D=\psi_2(R, R_B, P, F, X_F)$$

$$B=F-D$$

$$X_B=FX_F-DX_D$$

The X values may refer to a key component or to a split between light and heavy components. This system gives 5 degrees of freedom, such as F, X_F, P, and 2 of

X_D, D, B, X_B, R, R_B (but not D and B, nor $X_D D$ and $X_B B$).

Williams-Otto Plant Example

To study a process containing a number of units, the Williams-Otto hypothetical plant (8), Figure 5, is a good example. It is desirable to use unit by unit sets of equations rather than the overall balances. For this reason, the Williams-Otto equations can be rearranged to the following sets of equations.

REACTIONS

$$A+B\rightarrow C$$

$$C+B\rightarrow P+E$$

$$P+C\rightarrow G$$

EQUATIONS (Material Balances)

$$F_{RP}-0.1\,F_{RE}-F_P=0$$

$$(M_E/M_B)k_2\,F_{RB}F_{RC}V\rho/F_R^2-F_D F_{RE}/(F_R-F_G-F_P)=0$$

$$[k_2\,F_{RB}F_{RC}-(M_P/M_C)k_3\,F_{RC}F_{RP}]$$
$$\cdot V\rho/F_R^2-F_D(F_{RP}-F_P)/(F_R-F_G-F_P)-F_P=0$$

$$F_A-k_1\,F_{RA}F_{RB}V\rho/F_R^2-F_D F_{RA}/(F_R-F_G-F_P)=0$$

$$F_B-(k_1\,F_{RA}F_{RB}+k_2\,F_{RB}F_{RC})$$
$$\cdot V\rho/F_R^2-F_D F_{RB}/(F_R-F_P-F_G)=0$$

$$[(M_C/M_B)k_1\,F_{RA}F_{RB}-(M_C/M_B)k_2\,F_{RB}F_{RC}$$
$$-k_3\,F_{RP}F_{RC}]V\rho/F_R^2-F_D F_{RC}/(F_R-F_G-F_P)=0$$

$$(M_G/M_C)k_3\,F_{RC}F_{RP}V\rho/F_R^2-F_G=0$$

$$F_{RA}+F_{RB}+F_{RC}+F_{RE}+F_G+F_{RP}-F_R=0$$

and k_1, k_2, k_3 are functions of T only as follows:

$$k_1=5.9755(10^9)e^{-12000/T}$$

$$k_2=2.5962(10^{12})e^{-15000/T}$$

$$k_3=9.6283(10^{15})e^{-20000/T}$$

where $580°R < T < 680°R$, $\rho = 50$ lb/ft^3, F units = 1 lb/h, V = volume ft^3, $M_A = 100$, $M_B = 100$, $M_C = 200$, $M_E = 200$, $M_P = 100$, $M_G = 300$ and F_P desired = 4763 lb/h (depends on availability of A).

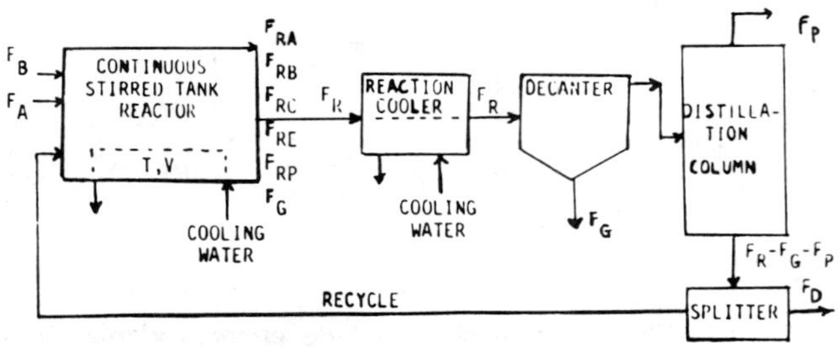

Figure 5. Williams-Otto plant.

Reactor:

Balance on E:

$$\frac{M_E}{M_B}k_2(T)F_{RB}F_{RC}\frac{V\rho}{F_R^2}+F_{RRE}=F_{RE}$$

where k_2 is a function of T only, V is fixed in operation and control, and F_{RRE} is the flow of E in the recycle stream.

Balance on P:

$$\left[k_2(T)F_{RB}F_{RC}-\frac{M_P}{M_C}k_3(T)F_{RP}F_{RC}\right]\frac{V\rho}{F_R^2}+F_{RRP}=F_{RP}$$

Balance on A:

$$F_A-k_1(T)F_{RA}F_{RB}\frac{V\rho}{F_R^2}+F_{RRA}=F_{RA}$$

Balance on B:

$$F_B-[k_1(T)F_{RA}F_{RB}+k_2(T)F_{RB}F_{RC}]\frac{V\rho}{F_R^2}+F_{RRB}=F_{RB}$$

Balance on C:

$$\left[\frac{M_C}{M_B}k_1(T)F_{RA}F_{RB}-\frac{M_C}{M_B}k_2(T)F_{RB}F_{RC}-k_3(T)F_{RP}F_{RC}\right]$$

and

$$\frac{V\rho}{F_R^2}+F_{RRC}=F_{RC}$$

Balance on G:

$$\frac{M_G}{M_C}\left[k_3(T)F_{RC}F_{RP}\frac{V\rho}{F_R^2}\right]=F_{RG}$$

Assuming G in recycle = 0,

$$F_R=F_{RA}+F_{RB}+F_{RC}+F_{RE}+F_{RG}+F_{RP}$$

$$T=f(F_{HX}, F_R, F_{RA}, F_{RB})$$

where F_{HX} is flow of heat exchange fluid.

Degrees of Freedom in Reactor:

Variables:

F_A, F_B, T, F_{RA}, F_{RB}, F_{RC}, F_{RE}, F_{RP}, F_{RG}, F_R, F_{RRA}, F_{RRB}, F_{RRC}, F_{RRE}, F_{RRP}, F_{HX}

$$N_F=N_V-N_E=16-8=8$$

Reaction Cooler:

$$T_C=f(F_R, T, F_{CW})$$

Degrees of freedom are

$$N_F=N_V-N_E=4-1=3$$

where F_{CW} is the flow of cooling water.

Decanter:

$F_G = f(F_R, T_C, F_{RG})$

$N_F = N_V - N_E = 4 - 1 = 3$

Distillation Column:

$F_p = f(F_{RP}, R, R_B, P, F_R - F_G)$

$P_Q = f(F_{RP}, R, R_B, P, F_R - F_G)$

$B = F_R - F_G - F_P$ (B is not necessary)

$F_{BP} = F_{RP} - F_P$ (F_{BP} is not necessary)

where P_Q = product quality and should equal 100% P. F_P should be 4763 lb/h, R is the reflux and R_B is the reboiler rate.

$N_F = N_V - N_E = 8 - 2 = 6$ (or $10 - 4 = 6$)

Splitter:

$F_D = \phi(F_R - F_P - F_G)$

$F_{RRA} = F_{RA}(1-\phi)$

$F_{RRB} = F_{RB}(1-\phi)$

$F_{RRC} = F_{RC}(1-\phi)$

$F_{RRE} = F_{RE}(1-\phi)$

$F_{RRP} = (F_{RP} - F_P)(1-\phi)$

Assuming F_P is 100% P,

$N_F = 15 - 6 = 9$

These equations assume steady-state and do not include hold-up quantities. Steady-state controls should be applied in addition to control of the necessary variables in the above equations.

The blocks can then be represented with their interrelationships as shown in Figure 6. The total degrees of freedom for the whole system can be obtained from the sum of the degrees of freedom of all blocks, $8+3+3+6+9=29$, minus the number of times variables appear in more than 1 block. Since T appears in two blocks, subtract one, since F_R appears in all five blocks, subtract four, and for F_{RG} subtract one, for F_{RP} and F_G two, for F_{RE}, F_{RC}, F_{RB}, F_{RA}, F_p, F_{RRP}, F_{RRE}, F_{RRC}, F_{RRA}, and T_C, subtract one each. The total degrees of freedom are then $29-21=8$ degrees of freedom, and the problem is to control eight or less of the variables to insure proper performance.

Considering Figure 6, if F_A is fixed by a previous unit, F_A (as far as this system is concerned) would be independent, leaving 7 degrees of freedom. The overall rate would be determined by F_A, and F_B (or F_B/F_A) could be controlled appropriately. Product quality should also be controlled, for example by top temperature control in the column, manipulating reflux (with a valve for a complete condenser or cooling water for a partial condenser). These three variables become inputs and R becomes an output. To insure appropriate reaction conditions, T should be controlled by manipulating F_{HX}, with T going in both units and F_{HX} an output. Similarly, T_C should be controlled for appropriate decanting, with F_{CW} an output. This leaves three more variables to control. At such a point it may be desirable to consider other variables and their probable direction. F_R, F_{RP}, F_{RG}, F_{RE}, F_{RC}, F_{RB}, and F_{RA} should probably leave the reactor and enter other units, if they are not controlled. All F_{RR} quantities should probably leave the splitter and enter the reactor. If these variables are given a direction, then the reactor, cooler, and decanter have inputs equal in number to their degrees of freedom, and no more variables in these units can be inputs. F_G should probably be an output from the decanter and an input variable to the splitter and distillation. If these directions are fixed, the three remaining possible control variables are now limited to P, R_B, F_p, F_D and ϕ. P, distillation pressure, should be fixed by control or otherwise, and for steady-state operation R_B could be controlled (by steam flow control). If R_B and P are inputs, the distillation operation is fixed and F_p must be an output and an input variable to the splitter. Then either F_D or ϕ could be the remaining possible control variable, and ϕ would probably be more appropriate for variations in F_A. These results are shown in Figure 7, with fixed (F) and control (C) variables indicated. Figure 8 shows the flow diagram with controls.

For steady-state control, level controllers manipulating outflows could be used in the reactor, decanter,

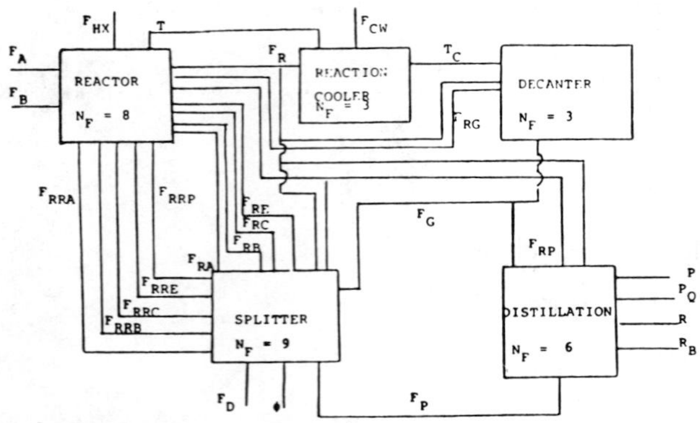

Figure 6. Block diagram of variables and degrees of freedom.

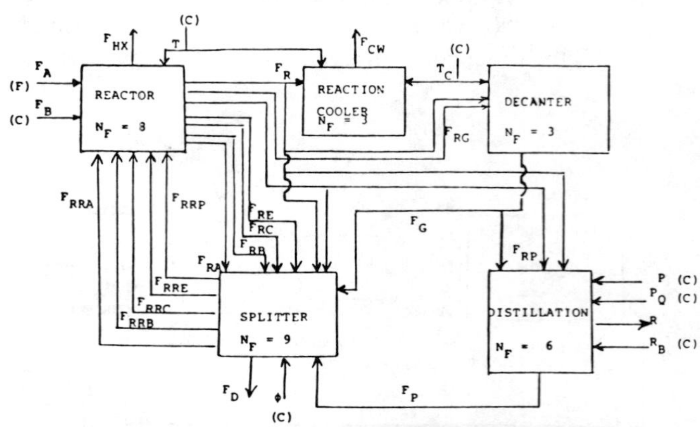

Figure 7. Block diagram showing independent variables, fixed (F) and controlled (C).

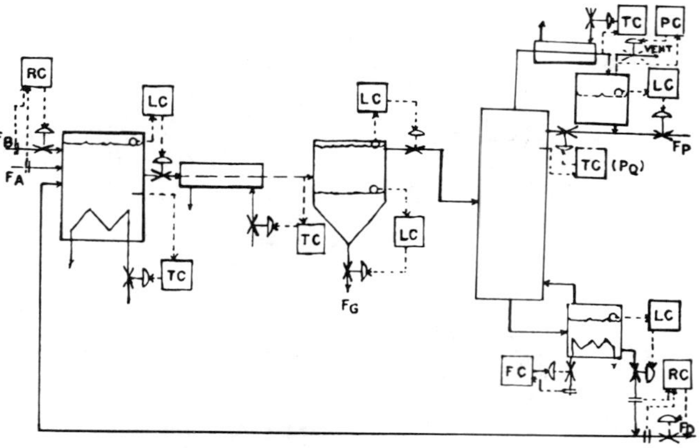

Figure 8. William-Otto plant with controls; where RC = ratio controller, LC = level controller, TC = temperature controller, and PC and FC stand for pressure and flow controllers, respectively.

distillate tank, and reboiler. The decanter would need two, one at the top and one at the interface separating F_G from F_R.

To include the Williams-Otto Plant as part of a larger complex after an initial analysis, the whole system could be considered as a block only showing variables connected with other processes, such as F_A, F_B, F_D, F_P, P_Q, F_G with degrees of freedom equal to the number of these variables selected to be fixed or controllable, in this case, three.

DISCUSSION

The advantage of this type of analysis and selection is that both the individual units and the overall process are given consideration more or less simultaneously. It also prevents the selection of control variables on an overall basis which are not possible on a unit basis. A very large system could be broken down into subsystems, and each subsystem could be tentatively analyzed, after which the overall system could be analyzed in terms of the subsystems.

NOMENCLATURE

B = bottoms product flow
C = number of components
D = distillate product flow
F = feed flow
H = enthalpy
L = liquid flow mol
N = number of distillation stages
N_E = number of equations
N_F = number of degrees of freedom
N_i = molar flow of component i
N_V = number of variables
P = pressure
Q = heat flow
R = reflux
R_B = reboil rate
T = temperature
V = vapor flow mol
W_S = steam flow
X = mol fraction liquid
Y = mol fraction vapor
F_{RA}, R_{RB}, F_{RC}, R_{RE}, F_{RP}, F_G, F_R, F_A, F_B, F_P, T, V, F_P, T_C, F_{HX}, F_{CW}, F_{RG}, F_{RRA}, F_{RRB}, F_{RRC}, F_{RRE}, F_{RRP}, Q, P_Q = Williams-Otto Plant Variables

LITERATURE CITED

1. Buckley, P. S., "Techniques of Process Control," 98, Wiley & Sons, New York, (1964).
2. Coughanowr, D. R., and L. B. Koppel, "Process Systems Analysis and Control," McGraw-Hill, New York (1965).
3. Douglas, J. M., "Process Dynamics and Control," Vol. 2, Prentice-Hill, Englewood Cliffs, New Jersey (1972).
4. Harriott, P., "Process Control," McGraw-Hill, New York (1964).
5. Hougen, J. O., and N. G. Brockmeier, "Instrumentation Technology," (Aug. 1969) (Oct. 1969).
6. Perry and Chilton, Kirkpatrick, ed., "Chemical Engineers' Handbook," McGraw-hill, New York (1963).
7. Rudd, D. F., and C. C. Watson, "Strategy of Process Engineering," 56, Wiley & Sons, New York (1968).
8. Williams, T. I. and R. E. Otto, *Trans. Am. Inst. Elect. Engr.*, 79, 458-473 (Nov., 1960).

SUGGESTED COMPLEMENTARY READING

1. "Chemical Engineers' Handbook", 5th Ed., Pages 22–28 to 22–30, 22–118 to 22–138. McGraw-Hill, N.Y. (1973). See Ref. (*6*).
2. Buckley, P. S., "Techniques of Process Control," Ch. 13, Wiley, N.Y. (1964).

STUDY PROBLEMS

1. In a particular mixing system, the inflow has a flow controller and the open mixing tank has a level controller manipulating outflow. A downstream operator wants to put a flow controller on the outflow. What would happen if this was done, the level controller was disconnected, and the inflow controller happened to be 0.1% lower than the outflow?

2. What would be a better way to hold the outflow and level constant and inflow approximately constant in the case of Study Problem 1?

3. A closed liquid-filled tank has one inflow, F_I, and two outflows, F_{OA} and F_{OB}. Considering the total flows at steady state, how many degrees of freedom are there? Assume all three lines have flow controllers, manipulating valves in their lines, what would happen if the controller setpoint for F_I was greater than the sum of setpoints for F_{OA} and F_{OB}?

4. What would happen in the situation described in Study Problem 3 if the F_I setpoint was less than the sum of F_{OA} and F_{OB} setpoints? Could one tell whether F_{OA} or F_{OB} would be in error?

5. Consider two tanks in series with level controllers and no side streams. Could the first be level-controlled by manipulating the outflow and the second by manipulating the inflow? Why or why not?

6. In the situation in Study Problem 5, how would one control levels and flows if:
   a) the inflow to tank 1 was critical?
   b) the inflow to tank 2 was critical?
   c) the outflow from tank 2 was critical?

7. In a flash separator at a fixed temperature and pressure with a binary solution, the following independent equations exist:

   1. $F=D+B$
   2. $FX_F=DX_D+BX_B$
   3. $X_D=K_1X_B$
   4. $1-X_D=K_2(1-X_B)$

   The number of degrees of freedom is $6-4=2$ (K_1 and K_2 are fixed constants). Can both X_D and X_B be controlled? Why or why not?

8. In the system in Problem 7, can one control X_D? Check this by using a block diagram with one block for each equation.

9. For the system in Study Problem 7, assume pressure and temperature are variable with the following additional equations, with K_1, K_2, T and P as additional variables.

   5. $K_1=f_1(T, P)$
   6. $K_2=f_2(T, P)$

   There are now four degrees of freedom, if F is fixed, three can be independent. Show why one cannot control D, and B at steady state.

10. Again using the system in Study Problems 9 and 7, why can't DX_D be greater than FX_F? Does this limitation show up in analyzing the degrees of freedom? How, or why not?

HOMEWORK PROBLEMS

1. A double effect evaporator can be represented by two evaporators as shown in Figure 9. Pressure in evaporator 2 is fixed at 1 atm by the pressure of the outlet system, but the pressure in the first evaporator must be above atmospheric pressure for the second evaporator to be effective, and this pressure can be varied. The feed rate may be varied, but the feed composition is fixed in previous processing. Devise a control system to maintain X_{B2} and steady state with steady operating conditions. The following equations apply.*

   Both Evaporators

   $$F=B+D$$

   $$FX_F=BX_B+DX_D$$

   $$X_D=0$$

*Steam supply does not limit heat transfer. Also, $W_{s2}=D_1$, $F_2=B_1$, possibly others.

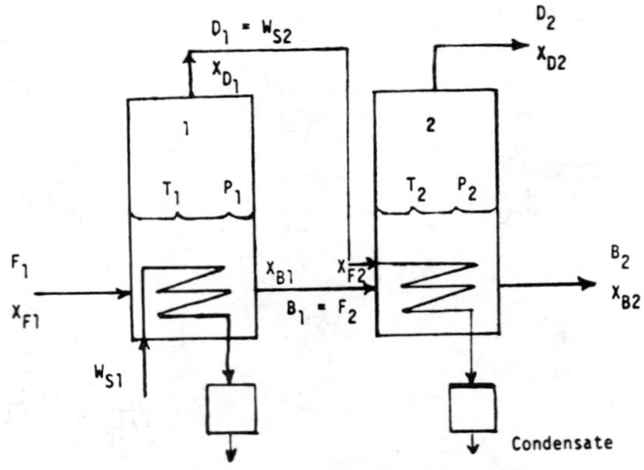

Figure 9. Double-effect evaporator.

   Evaporator 2

   $$D_2=\psi_1(T_2\, W_{s2}, F_2, P_1)$$
   (Heat Balance)

   $$P_1=\psi_2(T_2, W_{s2})$$
   (Heat Transfer)

   $$T_2=\psi_3(P_2, X_{B2})$$

   Evaporator 1

   $$D_1=\psi_4(T_1, F_1, W_{s1}, P_1)$$
   (Heat Balance)

   $$T_1=\psi_4(P_1, X_{B1})$$

2. A fixed flow and fixed temperature reaction product stream must be cooled down to T_3 crystallize out a by-product, and then be reheated to T_5 for further processing. To conserve energy (coolant and heating fluids), a recovery heat exchanger is to be installed, possibly according to Figure 10. The crystallizer operates full and closed, but the concentrated crystal suspension level can be measured and should be maintained well above the outlet. The fraction (Xs) of suspension to total flow is fixed by the previous reaction, but is not necessarily constant. Some of the equations involved are as follows:

   Heat Exchanger No. 1:

   $T_2=\psi_1(F, T_1, F_2, T_3)$ Heat Transfer

   $T_4=\psi_2(F, T_1, T_2, F_2, T_3)$ Heat Balance

   Heat Exchanger No. 2:

   $T_3=\psi_3(F, T_2, F_C)$

   Heat Exchanger No. 3:

   $T_5=\psi_4(F_2, T_4, F_H)$

   Devise a control system for this potential heat recovery system.

3. An alternative to the system in Homework Problem 2 is possible because T_5 is less than T_1, and could be obtained without heat exchanger 3 (with $T_4=T_5$).

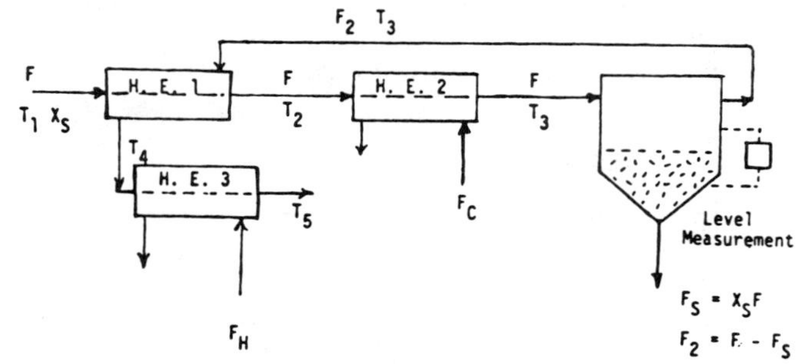

Figure 10. Crystallizer with heat recovery.

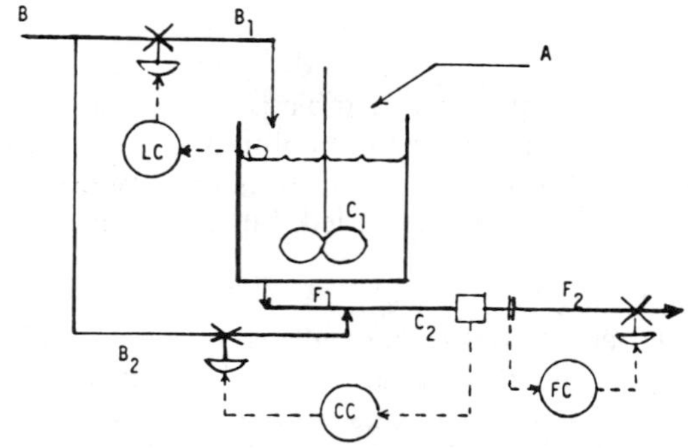

Figure 11. Suspension control system.

T_3 is not as critical as T_5 (T_4 in this case), and it may be possible to allow T_3 to be dependent on the normal operating conditions. Devise a control system if heat exchanger 3 is eliminated from the design.

4. A third possibility, based on Homework Problems 2 and 3, is that heat exchanger 3 will be eliminated, but T_3 may eventually require controlling along with T_4. All heat exchangers will have sufficient excess capacity to allow some physical flexibility in methods of operation and reduction in heat transfer capability. Can you devise a contingency system to control both T_3 and T_4, if necessary with minor process modifications?

5. A solid, A is suspended in a liquid, B, in the process shown in Figure 11, with the existing control system shown. Both the concentration of A, C_2, and the flow F_2 are important to the following reaction and process, but C is more important. Due to storage and handling problems the rate of addition of A is manually controlled along with the set point of the F_2 flow controller, partially based on the following process requirements. However adjustments are made infrequently compared to the hold-up time in the tank. Thus the rate of A fed and F_2 set point should be considered fixed. Control has been inadequate and modifications are needed. Someone has suggested that additional B_2 flow causes a decrease in B_1 flow and accomplishes little. Another suggestion is that the overall solids balance limits flexibility. Some of the equations are:

Independent Material Balances:

$$A + B_1 = F_1$$

$$C_1 = \frac{A}{F_1}$$

$$F_1 C_1 = F_2 C_2$$

$$F_2 = F_1 + B_2$$

Overall Balance on Solids:

$$A = F_2 C_2$$

(Derivable from Independent Balances)

Analyze this system for problems, if necessary, by considering all limitations, and see if you can determine the problem and devise a system to maintain steady state, and C_2, and also F_2, to the extent possible with manual control of A.

6. A distillation system consists of 2 columns to separate volatile components from heat degradable, less volatile products. The first column operates at fixed atmospheric pressure and the second operates under vacuum. Since the first column overhead vapor is at a higher temperature than the second reboiler, heat, cooling water, and also heat exchange area can be saved by condensing part of the first column vapor in the second column's reboiler, providing sufficient heat for the second column. The proposed system is shown in Figure 12, and your assignment is to develop a simple control scheme for further study and modification. The operating supervisor has warned that tying systems like this together causes severe control and operating problems and he will not approve any system with connecting streams that do not have fixed flow controllers to prevent fluctuations in one column from affecting the other. He also wants the temperatures in the condensate streams to the distillate drums controlled (complete condensation). Each

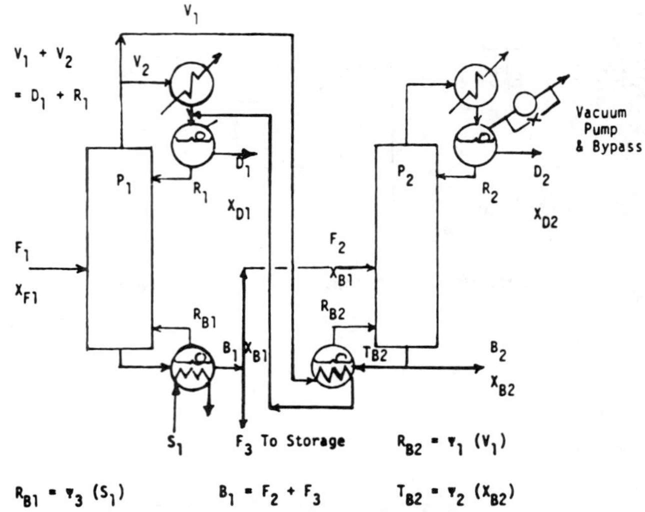

Figure 12. Distillation system with heat recovery.

column can be considered to be a simplified distillation block as discussed in the text, modified by variables V_1, V_2, F_3, and T_{B2} and their relationships. An important variable is the temperature of the reboiler in the second column at a specified vacuum. All inputs and outputs have storage facilities and may be varied, but input compositions are fixed.

7. A two-stage countercurrent extraction system is being designed to recover organic acids and reduce environmental problems from a waste stream as shown in Figure 13. The overflows are to act as level controllers on all four tanks. It is desired to maintain a certain concentration, Y_{A2}, in the exit solvent stream for economic recovery. The best way to do this is felt to be by manipulating solvent flows, since the waste stream flow and concentration may vary, as fixed by the waste generating process. Devise a control system for the equipment as shown in the figure, and discuss the dynamics of the major control loop. If you feel there might be problems with this system, discuss the possibility of equipment revisions to improve control.

8. A portion of a plant is being converted to produce benzene from toluene, and requires a new control system. The system and the most important variables and equations are shown in Figure 14, and Table 2. It is desired to maintain steady operation of flows, pressures and temperatures, especially at reactor inlet and flash tank to prevent decomposition and unwanted by-products. Concentrations of benzene and toluene are not critical at this point. The primary feed stream (1) is almost pure toluene and is fixed previously. A secondary feed from storage (2) mostly toluene, can be used to provide constant feed rates. A quench stream (14) is used to cool the product, by mixing, down to a desired temperature. Since this stream recirculates to the flash tank it does not involve a material balance and its desired flow depends on the desired heat balance. Select a control system.

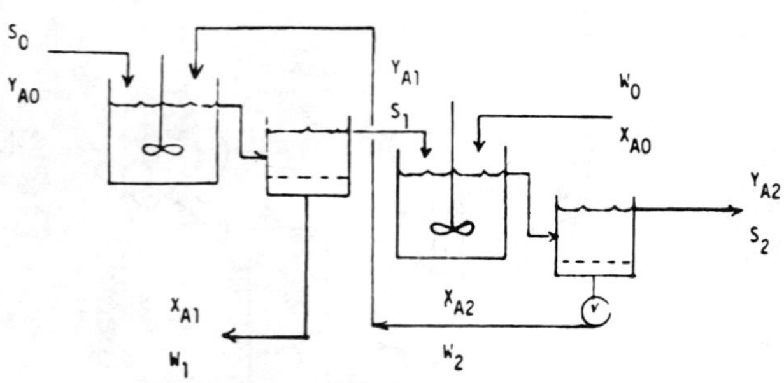

Figure 13. Countercurrent extraction system.

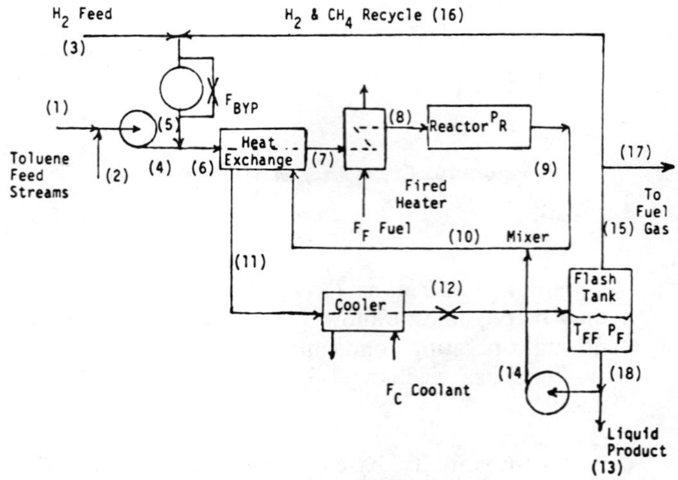

Figure 14. Portion of Benzene plant.

Table 2. Equations relating important variables in the process. T = Toluene, B = Benzene, M = Methane, H = Hydrogen, F = Total Flow, all in mol/h. Subscripts refer to stream numbers. P = Pressure, T_F = Temperature.

Pump, Compressor, Mixer

1. $T_1+T_2=T_4$
2. $H_3+H_{16}=H_6$
3. $H_6/(B_2+T_4)=R_H$
4. $H_6/M_{16}=\psi_1(F_{BYP})$
5. $F_6=B_2+T_4+H_6+M_{16}$
6. $F_4=B_2+T_4$
7. $B_2/T_2=X_B$

$N_F=13-7=6$

Reactor

1. $X=\psi_5(T_{F8}, F_6, T_4, P_R)$
2. $T_{F9}=\psi_6(X, F_6, T_{F8})$
3. $H_9=H_6-X$
4. $M_9=M_{16}+X$
5. $T_9=T_4-X$
6. $B_9=B_2+X$
7. $F_9=F_6$

$N_F=14-7=7$

Cooler

1. $T_{F12}=\psi_9(F_C, T_{F11}, F_{10})$

$N_F=4-1=3$

Heat Exchange

1. $T_{F7}=\psi_2(F_6, T_{F10}, F_{10}, T_{F11})$
2. $T_{F11}=\psi_3(F_6, T_{F10}, F_{10}, T_{F7})$

(Heat Transfer & Heat Balance)

$N_F=5-2=3$

Fired Heater

1. $T_{F8}-T_{F7}=\psi_4(F_F, F_6)$

$N_F=4-1=3$

Flash & Splitter

1. $T_{FF}=\psi_7(T_{F12}, P_F, T_9, B_9, M_9, H_9)$
2. $H_{15}=H_9$
3. $M_{13}=\psi_8(T_{FF}, M_9, F_9, P_F)$
4. $B_{13}=B_9$
5. $T_{13}=T_9$
6. $M_{15}=M_9-M_{13}$
7. $F_{15}=M_{15}+H_{15}$
8. $H_{16}=H_{15}(1-F_{17}/F_{15})$
9. $M_{16}=M_{15}(1-F_{17}/F_{15})$

$N_F=17-9=8$

Mixer

1. $T_{F10}(F_{10})=T_{F9}F_9+T_{FF}F_{14}$
2. $F_{10}=F_9+F_{14}$

$N_F=6-2=4$

APPENDIX

Solutions to the Study Problems

Solutions to the Homework Problems are available as a separate reprint from the AIChE Educational Services Dept., 345 East 47th St., New York, NY 10017.

SOLUTIONS TO THE STUDY PROBLEMS

1. Pulse testing gives similar results to sinusoidal testing, without the time and difficulty. Unlike step testing which excites only low frequencies in a system, pulse testing can determine dynamic behavior over the higher frequencies as well. Among the benefits derived from frequency response testing is the ability to identify characteristics of system components. If the test system is an on-line manufacturing process, pulse testing will avoid interruption of production and product quality.

2. The normalized frequency content of an input pulse is an indication of the ability of the pulse to excite the test system over the range of frequencies, and therefore gives a measure of the reliability of test results over the range of frequencies.

3. If the input pulse height is to high it may drive the system into its non-linear range, and our analysis assumes linearization. If pulse height is too low, it may fail to excite the system adequately. If pulse width is too great, higher frequencies may not be excited; and if width is too narrow, the pulse may not drive the system sufficiently.

4. Output pulse may be sampled periodically. To ensure reliable results at high frequencies, the accuracy criteria should be used to select the sampling interval: $(\omega \Delta t) = \pi/10$

5. Pulse tests can be applied to non-linear systems, which includes most systems encountered in practice. Linearity may be assumed when perturbations do not deviate greatly fron the steady-state operating point.

6. Truncation in this particular case should be avoided, because it will definitely cause erroneous results.

7. First, attempt to eliminate the source of the noise if possible, and be sure all test instruments and lead wires are adequately shielded fram electrical interference. Second, be sure that the input pulse form is exciting the system sufficiently in the frequency range of the noise. Check the normalized frequency content of the input pulse, which may be determined from the computer output by dividing the magnitude of the input pulse transform vector by the zero-frequency magnitude of the same. Adjust pulse height and width in future tests to improve excitation.

8. The computer program, PULSE, provides two means for correcting an open-ended pulse. One uses a correction algorithm, the other uses a linear base-line through the origin and the final point.

9. Pure delay time would not cause erroneous results, but would increase the phase difference proportionately with increasing frequency. The amplitude ratio would not be affected in any way. The effects of dead-time could be eliminated from the Bode diagram by not starting the output pulse until it first deviated from steady-state, i.e. subtracting the dead-time out at the start.

10. If both pulses are identical, a one-to-one dynamic ratio is implied. If this is not realistic, then perhaps the test did not excite the higher frequencies sufficiently. If the output pulse is delayed in time, pure delay time is indicated, i.e. transportation delay or plug flow may be involved.

SOLUTIONS TO THE STUDY PROBLEMS

(1) $a_{11} = \frac{\partial f_1}{\partial P_1} = k_o(DV[1]) \frac{\partial \phi_0}{\partial P_1} - k_1(DV[2] \frac{\partial \phi_1}{\partial P_1}$,

$a_{12} = \frac{\partial f_1}{\partial P_2} = - k_1(DV[2]) \frac{\partial \phi_1}{\partial P_2}$, (at critical flow $a_{12} = 0$),

$a_{21} = \frac{\partial f_2}{\partial P_1} = k_1(DV[2]) \frac{\partial \phi_1}{\partial P_2}$

$a_{22} = \frac{\partial f_2}{\partial P_2} = k_1(DV[2]) \frac{\partial \phi_1}{\partial P_2} - k_2 \frac{\partial \phi_2}{\partial P_2}$

$b_{11} = \frac{\partial f_1}{\partial (DV[1])} = \phi_0(P_1,P_0) \cdot \frac{\partial k_0}{\partial (DV[1])}$

$b_{12} = \frac{\partial f_1}{\partial (DV[2])} = -\phi_1(P_1,P_2) \cdot \frac{\partial k_1}{\partial (DV[2])}$

$b_{14} = \frac{\partial f_1}{\partial P_0} = k_0(DV[1]) \cdot \frac{\partial \phi_0}{\partial P_0}$

$b_{22} = \frac{\partial f_2}{\partial (DV[2])} = \phi_1(P_1,P_2) \frac{\partial k_1}{\partial (DV[2])} = - b_{12}$

$b_{23} = \frac{\partial f_2}{\partial k_2} = - \phi_2(P_2,P_3)$

$$b_{25} = \frac{\partial f_2}{\partial P_3} = -k_2 \frac{\partial \phi_2}{\partial P_3}$$

$$b_{13} = \frac{\partial f_1}{\partial k_2} = 0 \; ; \; b_{15} = \frac{\partial f_1}{\partial P_3} = 0 \; ; \; b_{21} = \frac{\partial f_2}{\partial (DV[1])} = 0 \; ; \; b_{24} = \frac{\partial f_2}{\partial P_0} = 0.$$

where

$$f_1 = k_0(DV[1]) \cdot \phi_0(P_1,P_0) - k_1(DV[2] \cdot \phi_1(P_1,P_2)$$

$$f_2 = k_1(DV[2]) \cdot \phi_1(P_1,P_2) - k_2 \, \phi_2(P_2,P_3)$$

SOLUTIONS TO THE STUDY PROBLEMS

1. Advantages of feedforward control

   The FF controller takes corrective action before the process is upset, while FB control action occurs only after the process is upset.
   The FF controller does not affect the stability of an existing FB system.

   Disadvantages of feedforward control

   A reasonably accurate process model is required for FF control, while FB control is robust with respect to model uncertainty.
   FF control requires that the disturbance be measured, which may not be feasible.
   The ideal FF controller may be physically unrealizable, but approximations may provide satisfactory control.

2. In FF controller design, a model is an absolute requirement for implementation. The model can be either a physical model or a black box model. Often the steady state gain for the load transfer function can be determined from a steady state material or energy balance.

   A black box model can be developed experimentally for both the process transfer function and the load transfer function, by using standard step test or pulse test techniques. In most cases, however, lead/lag and feedforward gain units are iteratively "tuned" in the field until an "acceptable" response is obtained. This field tuning approach can use modeling to obtain a first guess at the setting but this is not always necessary.

3. Yes, the contributions of feedforward and feedback control systems are additive.

4. When feedforward control is combined with feedback control, the stability is not affected because the same characteristic equation is obtained with FF and FF + FB systems. This is because in the block diagram, the FF system is external to the FB loop.

5. A much better knowledge of the process is required for FF control, compared with FB control. However, a non-iterative (algebraic) design can achieve perfect control, while FB controller design requires several iterations to optimize the response.

6. FF control cannot be implemented physically if the process deadtime is greater than the load deadtime. In this case a predictor would be needed, which is unrealizable.

7. First order plus deadtime transfer functions are reasonable approximations for most chemical processes. Using first order plus deadtime models for the load and process transfer functions theoretically yields a lead/lag unit. Digital implementation gives us the capability to handle deadtime accurately, which is not possible with analog control.

SOLUTIONS TO THE STUDY PROBLEMS

1. The result follows from the equations given in the module.

2. The secondary measurement provides an early measurement and control actuation for disturbances entering before c_2.

3. This simply requires writing each role in terms of the variables.

4. The ratio between two variables.

SOLUTIONS TO THE STUDY PROBLEMS

1. Adaptive control is useful for processes with frequent changes in process dynamics, operating conditions, and/or environment. Adaptive control might be attractive in the following applications areas: control of a catalytic reactor when the catalyst loses activity, adjustment of flowrates in a heat exchanger to compensate for fouling, and adaptation of reflux in a distillation column in response to disturbances in the feed composition.
2. An example of programmed adaptation is the case of the once-through boiler. For this case, the tuning parameters are adjusted for changes in the feedwater flow rate by known relationships.
3. In order to implement self-tuning control, the system must

   1) estimate the process model parameters.

   2) calculate new controller parameters,

   3) implement the control.

   The model parameters are estimated either from input-output information or from deviations from a given reference model. The controller calculations are typically the solution of a set of simultaneous equations that relate the controller parameters to the process parameters. The implementation of control involves the calculation of the controller output based on the new parameters.

4. Continuous cycling of a process at its stability limit could be used in the process parameter estimation as the forcing signal of the estimation. This would allow use of tuning correlations based on critical frequency and period (e.g., Ziegler-Nichols). The cycling would almost surely be sufficient excitation for the parameter estimation, which is not guaranteed with white noise or PRBS signals. However, the cycling will not yield parameter estimates that are as statistically accurate as white noise or PRBS excitation. Also, letting the process oscillate defeats the purpose of control since a "steady" output is not obtained, and it is not desirable to operate close to the limits of stability.

5. The Leeds and Northrup controller is designed for a first order exponential rise to setpoint, while the self-tuning regulator is designed for return to setpoint in a fixed time, such as one sample interval for one-step-ahead control. The fixed time constraint for the STR can be too stringent for many processes. The L&N controller can be too stringent at times also, namely in cases where ringing occurs.

6. If overshoot occurs in the L&N controller, it must be the result of model errors, noise, or an improperly tuned controller (the L&N controller is designed for an exponential approach to setpoint with no overshoot). The occurrence of excessive model errors may require that the controller be placed in a non-adaptive mode because the parameter estimation is not working properly.

7. The L&N controller has potential problems with ringing and changes in time delays. The STR does not contain a parameter that can be altered to change the response characteristics, except for the sample time. As a result, the STR tends to be an "overactive" controller.

SOLUTIONS TO THE STUDY PROBLEMS

1. θ_D may include elements (a),(b),(f).
   θ_p may include elements (a),(b),(c),(d),(f).

2. θ_D has no effect on stability. The effect of θ_p is discussed in module.

3. Not necessarily. Too high a controller gain will cause instability.

4. See eqn.(8). Dynamic performance is always improved. The delay in the response to changes in set point remains unchanged. The delay in response to changes in load is extended in time:
   Originally:

$$\frac{Y(s)}{D(s)} = \frac{1}{1+G_cG_pe^{-\theta_ps}} \cdot G_De^{-\theta_Ds}$$

   Compensated:

$$\frac{Y(s)}{D(s)} = \frac{1}{1+C_cG_p} \cdot [(1+G_cG_p)G_De^{-\theta_Ds}-G_cG_pG_De^{-(\theta_D+\theta_p)s}]$$

SOLUTIONS TO THE STUDY PROBLEMS

1. The gain relating a manipulative input to the corresponding controlled variable should be high, while the gains relating that same input to other controlled variables should be negligible. However, constraints on manipulative inputs, constraints on controlled variables, noise, dynamic effects, and many other factors may also influence the choice.

2. The "loops" of classical feedback control are designed independently. Interactions appear as severe disturbance effects, since the effect is transmitted back and forth between the loops for some time. Interactions tend to degrade performance and can even lead to instability.

3. Even though the static process gains relating a manipulative input to a controlled variable might be very large, constraints on the input might limit the available effort so severely that it could not be used with that output.

4. The transpose of a matrix is a matrix with rows and columns interchanged.

5. One possible identity is $A^{-1} = \mathrm{adj}A/|A|$ where adj refers to the adjoint matrix and $|A|$ is the determinant of A.

6. The term $(\partial c_i/\partial m_j)_c$ refers to the static process gain relating a manipulative input m_j to a controlled variable c_i with all other controlled variables held constant. The term $(\partial c_i/\partial m_j)_m$ refers to the gain relating the same variables with all other manipulative inputs held constant.

SOLUTIONS TO THE STUDY PROBLEMS

1. With the inflow 0.1% lower than the outflow, the tank would gradually empty until either the flows were readjusted or the mixing tank went dry and the outflow equaled the inflow. At this point the outflow would equal the inflow, no matter how much the outflow control valve opened.

2. At steady state $F_{IN} = F_{OUT}$, N_F = 2 var. - 1 eq. = 1:

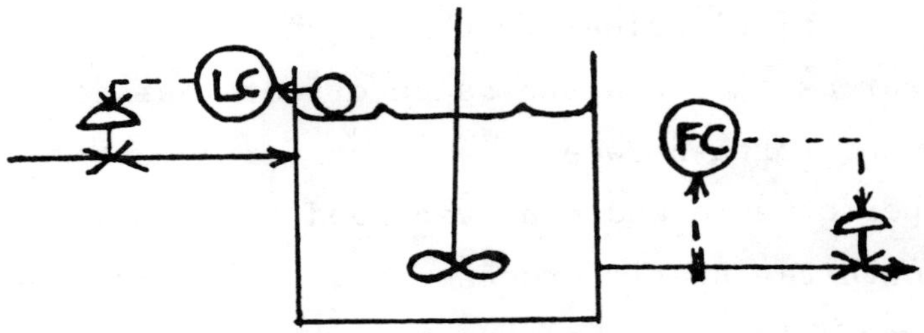

Use the 1 degree of freedom to control outflow. Use level control to maintain steady state by manipulating inflow.

3. Equations $F_I = F_{OA} + F_{OB}$ (steady-state)

$$N_F = 3 - 1 = 2$$

If F_I were set greater than $F_{OA} + F_{OB}$ then the flow in would be limited by the outflow, and no matter how much the F_I control valve opened the inflow would equal $F_{OA} + F_{OB}$.

4. If F_I were controlled lower than $F_{OA} + F_{OB}$ set points, the F_I valve would limit inflow and thus outflow and either or both F_{OA} and F_{OB} would be low. Either F_{OA} or F_{OB} could be most in error, depending on the fluid dynamic situation.

5.

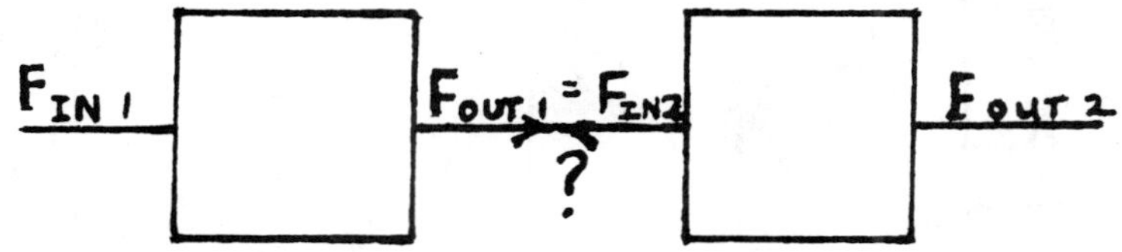

No, only 1 system can determine the one flow.

6.

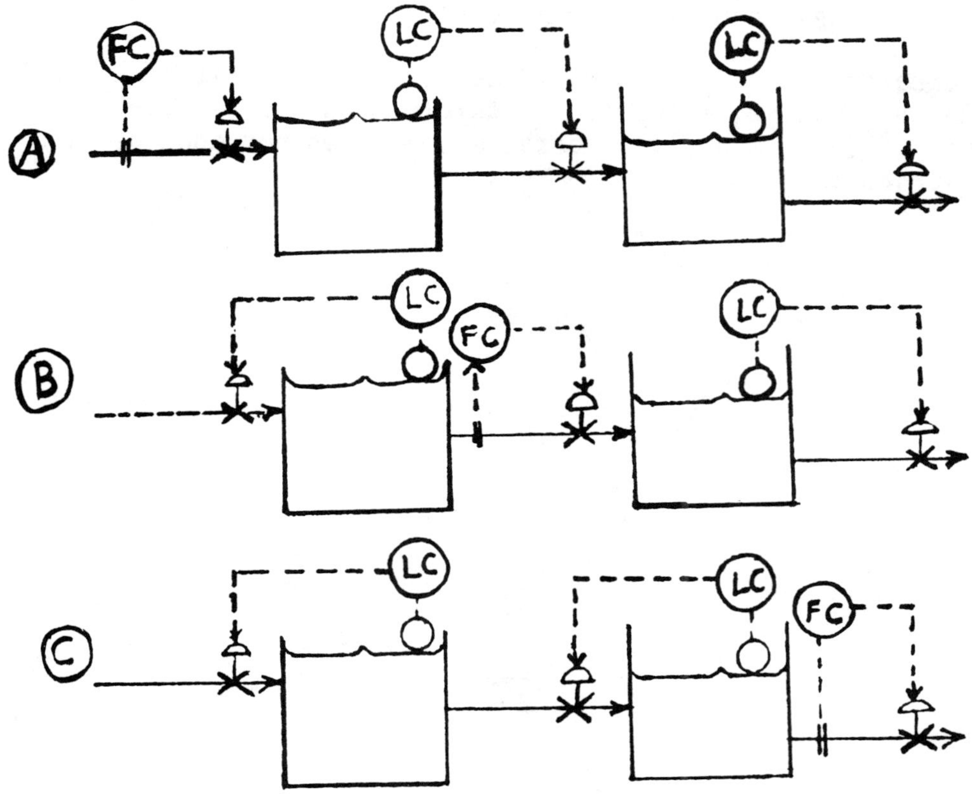

7. By inspection we cannot control both X_D and X_B because there is only 1 degree of freedom in Equation (3) thus only 1 of these could be controlled.

8. By inspection, there are only 2 variables in 2 equations, (3) & (4), so there are no degrees of freedom for these 2 variables.

   By diagram: If X_D = independent it is an input to 4 and 3. Since 3 and 4 have only 1 degree of freedom each, X_B must leave both, but a variable

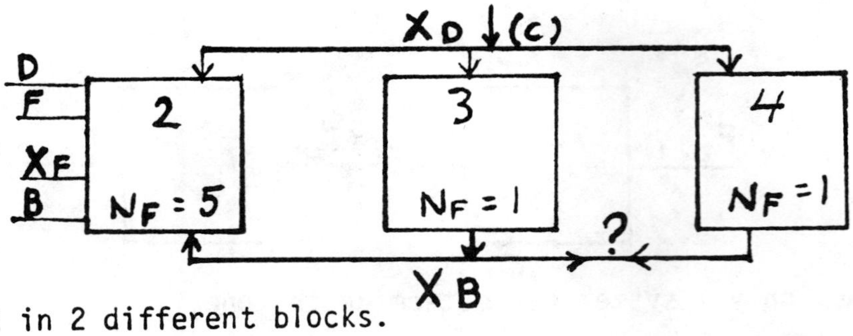

cannot be determined in 2 different blocks.

9. Equation 1 or Block 1 gives only 2 degrees of freedom among F, D, and B, so that only 2 of these can be controlled and/or fixed.

10. If DX_D were greater than FX_F, $X_B B$ would be negative and either X_B or B would have to be negative, which is impossible. This does not show up in the degrees of freedom analysis because this does not consider the sign of the variables.